Next-Generation Antennas

Scrivener Publishing
100 Cummings Center, Suite 541J
Beverly, MA 01915-6106

Advances in Antenna, Microwave, and Communication Engineering

Series Editors: Manoj Gupta, PhD, Pradeep Kumar, PhD

Scope: This book series represents an exciting forum for the presentation and discussion of the most recent advances in the antenna, microwave, and communication engineering area. In addition to scientific books, contributions on industrial applications are strongly encouraged, covering the above listed fields of applications. This book series is aimed to provide monograph, volumes, comprehensive handbooks and reference books that are empirical studies, theoretical and numerical analysis, and novel research findings for the benefit of graduate and postgraduate students, research scholars, hardware engineers, research and development scientists, and industry professional working towards the latest advances in antenna, microwave, and communication engineering and for their industrial applications.

Publishers at Scrivener
Martin Scrivener (martin@scrivenerpublishing.com)
Phillip Carmical (pcarmical@scrivenerpublishing.com)

Next-Generation Antennas

Advances and Challenges

Edited by
Prashant Ranjan,
Dharmendra Kumar Jhariya,
Manoj Gupta,
Krishna Kumar,
and
Pradeep Kumar

Scrivener
Publishing

WILEY

This edition first published 2021 by John Wiley & Sons, Inc., 111 River Street, Hoboken, NJ 07030, USA and Scrivener Publishing LLC, 100 Cummings Center, Suite 541J, Beverly, MA 01915, USA
© 2021 Scrivener Publishing LLC
For more information about Scrivener publications please visit www.scrivenerpublishing.com.

Wiley Global Headquarters
111 River Street, Hoboken, NJ 07030, USA

For details of our global editorial offices, customer services, and more information about Wiley products visit us at www.wiley.com.

Limit of Liability/Disclaimer of Warranty
While the publisher and authors have used their best efforts in preparing this work, they make no representations or warranties with respect to the accuracy or completeness of the contents of this work and specifically disclaim all warranties, including without limitation any implied warranties of merchantability or fitness for a particular purpose. No warranty may be created or extended by sales representatives, written sales materials, or promotional statements for this work. The fact that an organization, website, or product is referred to in this work as a citation and/or potential source of further information does not mean that the publisher and authors endorse the information or services the organization, website, or product may provide or recommendations it may make. This work is sold with the understanding that the publisher is not engaged in rendering professional services. The advice and strategies contained herein may not be suitable for your situation. You should consult with a specialist where appropriate. Neither the publisher nor authors shall be liable for any loss of profit or any other commercial damages, including but not limited to special, incidental, consequential, or other damages. Further, readers should be aware that websites listed in this work may have changed or disappeared between when this work was written and when it is read.

Library of Congress Cataloging-in-Publication Data

ISBN 9781119791867

Cover image: (Antenna Tower): Carmen Hauser | Dreamstime.com
Cover design by Kris Hackerott

Set in size of 11pt and Minion Pro by Manila Typesetting Company, Makati, Philippines

10 9 8 7 6 5 4 3 2 1

Contents

Preface

In the 21st century, the world is facing many challenges and developments. People moving into urban areas are keen to experience the new changes in cities, where facilities are more user-friendly and comfortable. It has led to the existence of next-generation antennas, and researchers in this field are working towards developing these antennas for industrial applications. Keeping this in view, the present book is aimed at exploring the various aspect of next-generation antennas, and their advances, along with their challenges, in detail.

Antenna design and wireless communication have recently witnessed their fastest growth period ever in history, and this trend is likely to continue for the foreseeable future. Due to recent advances in industrial applications as well as antenna, wireless communication and 5G, we are witnessing a variety of new technologies being developed. Compact and Low-cost antennas are increasing the demand for ultra-wide bandwidth in next-generation (5G) wireless communication systems and the Internet of Things (IoT). Enabling the next generation of high-frequency communication, various methods have been introduced to achieve reliable high data rate communication links and enhance the directivity of planar antennas. 5G technology can be used in many applications such as smart city and smartphones, and many other areas as well. This technology can also satisfy the fast rise in user and traffic capacity in mobile broadband communications.

Therefore, different planar antennas with intelligent beamforming capability play an important role in these areas. The purpose of this book is to present the advanced technology, developments, and challenges in antennas for next-generation antenna communication systems. This book is concerned with the advances in next-generation antenna design and application domain in all related areas. It includes a detailed overview of the cutting age developments and other emerging topics, and their applications in all engineering areas that have achieved great accuracy and performance with the help of the advances and challenges in next-generation antennas.

Readers

This book is useful for the Researchers, Academicians, R&D Organizations, and healthcare professionals working in the area of Antenna, 5G Communication, Wireless Communication, Digital hospital, and Intelligent Medicine.

The main features of the book are:

- It has covered all the latest developments and future aspects of antenna communication.
- Very useful for the new researchers and practitioners working in the field to quickly know the best performing methods.
- Provides knowledge on advanced technique, monitoring of the existing technologies and utilizing the spectrum in an efficient manner.
- Concisely written, lucid, comprehensive, application-based, graphical, schematics, and covers all aspects of antenna engineering.

Chapter Organization

Chapter 1 gives an overview of Microstrip filters for UWB communication. It also describes the Multiband Microwave filter, Ultra-Wideband (UWB) bandpass filter, and ultra-wideband filter with notch band characteristic.

Chapter 2 describes the introduction of 2×2 MIMO antenna configuration, and their diversity performance analysis.

Chapter 3 explains the Scilab open-source software and antenna array design.

Chapter 4 gives an overview of conformal antenna, explains characteristics of conformal antenna, wearable technology, cloth fabric wearable antennas, and simulated radiation pattern.

Chapter 5 gives an overview of On-Body wearable antenna for ISM band applications, explains design of star-shape with AMC backed structure, characterization of AMC unit cell, bending analysis of star-shaped antenna with AMC backed structure, and on-body placement analysis of the antenna with AMC structure.

Chapter 6 gives an overview of antenna miniaturization for IoT applications, issues in antenna miniaturization, antenna for IoT applications, and miniaturize reconfigurable antenna for IoT.

Chapter 7 gives an overview of wireless communication, Microstrip patch antenna, design & implementation of projected antenna, and observe the effect of different substrate materials.

Chapter 8 provides understanding of reconfigurable antenna for cognitive radio system, uses and drawbacks of reconfigurable antenna, and spectrum access and cognitive radio.

Chapter 9 describes the Ultra-Wideband filtering antenna, and Ultra-Wideband filtering antenna with notch band characteristic.

Chapter 10 describes the UWB and multiband reconfigurable antennas, need for reconfigurable antennas, triple notched band reconfigurable antenna, and tri-band reconfigurable monopole antenna.

Chapter 11 highlighted the IoT world communication through antenna propagation with emerging design analysis features, design and parameter analysis of multi-input multi-output antennas, measurement analysis in 3D pattern with IoT module.

Chapter 12 gives an overview of reconfigurable antennas, polarization reconfigurable antenna, compound reconfigurable antennas, and reconfigurable leaky wave antennas.

Chapter 13 gives an overview of design of compact Ultra-Wideband (UWB) antennas for microwave imaging applications, design of a UWB-based compact rectangular antenna, and validaed the miniaturized UWB antenna with the human breast model developed.

Chapter 14 gives an overview of joint transmit and receive MIMO beamforming in multiuser MIMO communications, and system modeling for MIMO beamforming architecture based on generalized least mean algorithm.

Chapter 15 describes the adaptive stochastic gradient equalizer design for multiuser MIMO system, and design of adaptive equalizer by minimizing BER.

.

1

Different Types of Microstrip Filters for UWB Communication

Prashant Ranjan[1]*, Krishna Kumar[2], Sachin Kumar Pal[3] and Rachna Shah[4]

[1]Department of ECE, University of Engineering and Management, Jaipur, India
[2]UJVN Ltd., Uttarakhand, India
[3]Bharat Sanchar Nigam Ltd., Guwahati, India
[4]National Informatics Centre, Dehradun, India

Abstract

Many filters such as triple-band filter, multiband filter, UWB filter, and notch band filters have been investigated in recent decades [1]. Bandpass filters with the features of good performance, micro-package, ease of use, and low cost have been the focus of device miniaturization. However, most of these UWB filters with band-notched have been designed by using various slots either in the ground plane or radiating patch, slit on feeding line, or integration of filter in feed line of the antenna. Slotted methods can be used for frequency rejection but it may distort the radiation patterns because of the electromagnetic leakage of these slots. In this chapter, a survey of multiband filter, UWB filter, and UWB with notch band filter are presented.

Keywords: Ultra-wideband, bandpass filter, microstrip patch, multiband, multiple-mode resonator, and transmission zeros

1.1 Introduction

The system can be streamlined and the physical dimension of the circuit minimized by triple-band microwave filters, thereby increasing the demand for triple-band microwave filters in modern communication systems. Recently, in many research papers, triple and multiband microwave

---

*Corresponding author: prashant.ranjan@uem.edu.in

Prashant Ranjan, Dharmendra Kumar Jhariya, Manoj Gupta, Krishna Kumar, and Pradeep Kumar (eds.) Next-Generation Antennas: Advances and Challenges, (1–22) © 2021 Scrivener Publishing LLC

filters have been widely studied. The use of alternately cascaded multiband resonators is one way of designing a triple-band filter. Coupling systems are used to achieve two and three frequency bands with Quasi-elliptic and Chebyshev frequency responses [2].

1.2   Previous Work

Various researchers have worked on the Microstrip filters for UWB communication.

1.2.1   Multiband Microwave Filter for a Wireless Communication System

Hao Di *et al.* [3] presented a technique to achieve a triple passband filter. In this method, a frequency transformation from the normalized frequency domain to the actual frequency domain is used. Applying this transformation, filter circuits with cross-coupling having triple-passband have been constructed. Cross coupled tri-band filter topology is presented, which consists of parallel resonators and admittance inverters. By using expressions, the external quality factors and coupling coefficients can be calculated. Three passbands 3.3–3.4, 3.5–3.6, and 3.7–3.8 GHz, with more than 20 dB return loss have been reported in this paper.

Hsu *et al.* [4] proposed asymmetric resonator-based one wideband and two tri-band BPFs. The resonator contains microstrip sections with different electrical lengths. Three resonant modes can be shifted to the desired center frequencies by varying the stub length of the first filter. In the second filter, asymmetrical resonators are used to achieve a wide stopband. For the third filter, wideband BPF is designed using multi-mode resonances and transmission zeros (TZs) of the asymmetrical resonator. It suppressed the higher-order harmonics. The three passbands 1.5, 2.5, and 3.5 GHz are achieved using four resonators.

Liou *et al.* [5], proposed a Marchandbalun filter with the shorted coupled line to achieve triple passbands. The filter is constructed with a triple-band resonator to exhibit the triple-band admittance inverter characteristic. The compensation techniques for phase–angle and impedance matching are used to improve the phase and amplitude responses of the existing three passbands. The defected ground structure stubs and microstrip coupled–line sections is used to realize the filter. Jing *et al.* [6] proposed a single multimode resonator-based filter with six passbands. The proposed MMR is a SIR (stepped impedance resonator) with two symmetrical open-circuited stubs positioned at two sides and one shorted stub connected in the middle. The electrical lengths of two open stubs are increased to excited the transmission zeros (TZs) and transmission poles (TPs). The TZs are separated from TPs by introducing open stubs; therefore, a six-band BPF is designed. Two input-output tapped branches and radical stub-loaded shorted lines are adjusted to improve filter performance.

Hong *et al.* [7] proposed a cross-coupled microstrip filter by using square open-loop resonators. In this paper coupling coefficients calculation of the three coupling structures of filters is developed. Empirical models are presented to estimate the coupling coefficients. A four-pole elliptic function type filter is designed. Three types of coupling characteristics,the electric, magnetic, and mixed couplings, have been reported. Kuo *et al.* [8] presented a microstrip filter with two frequency passband response based on SIR. SIR is in parallel-coupled and vertical–stacked configuration. Resonance characteristics of the second resonant frequency can be tuned over a wide range by adjusting its structure parameters. Tapped input/output couplings are used to match–band response for the two designated passbands. Both coupling length and gap are adjusted together to meet the required coupling coefficients of two bands. Fractional bandwidth design graphs are used to determine geometric parameters. Two passband resonant frequencies are 2.45 and 5.8 GHz with a fractional bandwidth of 12% and 7%, respectively. The measured insertion loss for the first passband is 1.8 dB and for the second passband is 3.0 dB. Higher-order filters are also designed using this design procedure.

Guan *et al.* [9] proposed a triple-band filter using two pairs of SIRs having single transmission zero. The first and the third frequency bands are realized by using Parallel coupled microstrip lines and the second frequency band is realized by using an end-coupled microstrip line. Single TZ is generated due to the antiparallel structure of the microstrip line. 2.4 GHz and 5.7 GHz are generated by longer resonators and 3.8 GHz is generated by the shorter resonators. By changing the impedance ratio of the resonator, the passband position of the filter can be adjusted. The filter bandwidths can be adjusted by adjusting the distance between resonators.

Ko *et al.* [10] presented two coupled line structures with open stubs to design a triple-band filter but insertion losses and bandwidths are poor. The third resonance frequency is shifted from 8.9 GHz to 6.5 GHz by variation in the lengths of two open stubs. A gap between transmission lines is used to adjust the second resonance frequency at 4.2 GHz. The first resonance frequency (2.4 GHz) can be adjusted using the coupled line length.

Lin *et al.* [11] proposed a triple-band BPF based on SIR. Hairpin type structure is used to reduce the size of the filter. Three passbands are 1.0,

2.4, and 3.6 GHz with an insertion loss of 2.2, 1.8, and 1.7 dB, respectively. Wibisonoet *et al.* [12] proposed a triple band BPF using cascaded three SIR. Filter passband frequencies are 900 MHz, 1800 MHz, and 2600 MHz simultaneously. Riana *et al.* [13] proposed a split-ring resonator to create three passband frequencies having two TZs. The filter coupling model approach is used to design filter and control passbands. Additional transmission zeros can be introduced by adjusting the position of the coupled resonators. Using triple-mode SRRs two filter topologies have been described. Using the lumped-element model to design the mainline and cross-couplings topology is presented in this paper. Two independent extraction methods, a de-tuning method, and a parameter-extraction method are used to determine coupling coefficients. Three passbands are 1.7, 2.4, and 3 GHz.

Liu *et al.* [14] proposed a triple band HTS (high temperature superconducting) filter using stub–loaded multimode resonator. The odd-even mode method is used to investigate the characteristics of the multimode resonator. A nonresonant node with a source-load coupling configuration is used to create TZs. Three passband resonant frequencies of the HTS filter, 2.45, 3.5, and 5.2 GHz, are presented. Insertion losses of the first, second, and third passbands are 0.16 dB, 0.55 dB, and 0.22 dB, respectively. The overall size of the filter is 8.3 mm × 8.6 mm. Qiang *et al.* [15] proposed a design of a wideband 90° phase shifter, which consists of open stub-based stepped impedance and a coupled–line to achieve wideband. The impedance ratio of the SIOS is used to analyze the bandwidths of return loss and the coupling strength of the coupled–line is used to analyze the phase deviation. The bandwidth of the phase shifter is 105% (0.75 to 2.4 GHz) with an insertion loss of 1.1dB.

Haiwen *et al.* [16] presented a triple band HTS filter by using a multimode stepped impedance split ring resonator (SI–SRR) to achieve the wide stopband property. Even and odd mode analysis is used to analyze the equivalent circuit model. This filter can be operated at 2 GHz, 3.8 GHz, and 5.5 GHz. The measured insertion losses can be obtained as approximately 0.19 dB, 0.17 dB, and 0.3 dB respectively at the center frequency of each passband. Zheng *et al.* [17] proposed a UWB BPF by creating triple notch-bands to make the multiband filter. SIR and four shorted stubs having a length of λ/4 are used to design the basic UWB filter. Open load stubs and E–shaped resonator are used to achieve triple band-notched performance. Three notched bands are at 4.8 GHz, 6.6 GHz, and 9.4 GHz. The minimum insertion loss of 0.6dB and a maximum ripple of 0.88 dB are reported.

Guan *et al.* [18] proposed a triple band HTS filter based on a coupled line SIR (C-SIR) to control transmission zeros. Three harmonic peaks are generated using C–SIR. Even–odd analysis method is applied to analyze the filter. An interdigital structure between the feed lines and C–SIR is used to increase the selectivity of the filter. Spiral-shaped lines are used for better coupling of the second-order resonator. Three frequency bands 1.57 GHz for GPS, 3.5 GHz for WiMAX, and 5.5 GHz for WLAN are achieved. Insertion losses are found to be 0.10, 0.20, and 0.66 dB at each passband, respectively.

Chen *et al.* [19] proposed multiband microstrip bandpass filters with circuit miniaturization. Five compact triple modes stub-load SIRs (SL–SIRs) are used to achieve five bands filter. The coupling scheme presented in this paper provides multiple paths for different frequency bands which gives more design flexibility. Centre frequencies of five bands are 0.6, 0.9, 1.2, 1.5, and 1.8 GHz. The insertion losses are approximately 2.8 dB, 2.9 dB, 2.9 dB, 2.6 dB, and 2.3 dB respectively. Wen *et al.* [20] proposed a six-band BPF based on semi-lumped resonators. The semi-lumped resonator included a chip inductor in the midpoint and two identical microstrip lines. Comparison between the semi-lumped resonator and conventional half-wavelength uniform resonator are presented. Harmonic frequencies are controlled by semi-lumped resonator. A distributed coupling technique is used to integrate bandpass filters. Low loading effects are achieved, which is essential for multiband circuits.

Wang *et al.* [23] proposed a compact UWB BPF having three notch bands by using a defected microstrip structure of U–shaped (UDMS). E–shaped MMR and interdigital coupled lines are used to obtain two transmission zeros at lower and upper passbands. The triple band-notched characteristics are achieved by introducing three parallel UDMSs. Notch band frequencies are 5.2, 5.8, and 8.0 GHz. A summary of previous work on multiband filters is given in the following Table 1.1.

1.2.2   Ultra-Wideband (UWB) Bandpass Filter

Wong *et al.* [27] proposed a UWB bandpass filter by using a quadruple–mode resonator. Two transmission zeros are generated by introducing two short-circuited stubs in MMR-based resonator. Two short-circuited stubs are used to control the fourth resonant mode and combining with the previous three resonant modes to make a quadruple–mode UWB filter. RT/Duroid 6010 substrate is used having a height of 0.635 mm, loss tangent 0.0023, and permittivity of 10.8. Interdigital coupled-lines are used to feed the MMR. Filter covers the frequency range of 2.8-11.0 GHz with a fractional bandwidth of 119%. Minimum insertion loss is found 1.1 dB within the UWB passband. Group delay Variation is found between 0.19 – 0.52 ns within UWB passband.

Table 1.1 Summary of multiband filters.

Sl. no.	Author	Year	Resonant frequencies	Technique used	No. of bands	Limitations
1	Chen et al. [2]	2006	2.3, 3.7, and 5.3 GHz	SIR	3	Poor insertion loss
2	Hao et al. [3]	2010	3.3, 3.5, and 3.7 GHz	Open-loop resonators	3	Poor selectivity
3	Chong et al. [5]	2013	2.1, 3.45, and 5.15 GHz	Coupled-Line Admittance Inverter	3	Very complex design
4	Jing et al. [6]	2016	0.7, 2, 3.2, 4.5, 5.8, and 7 GHz	MMR	6	Narrow bands
5	Jia et al. [7]	1996	2.46 GHz	Open-Loop Resonators	1	Poor insertion loss
6	Kuo et al. [8]	2005	2.45 and 5.8 GHz	SIR	2	Harmonics are present
7	Guan et al. [9]	2009	2.4, 3.8, and 5.7 GHz	SIR	3	Poor selectivity
8	Ko et al. [10]	2013	2.2, 4.2, 6.5, and 8.9 GHz	Open stubs	3	Narrow bands
9	Marjan et al. [21]	2006	2.65, 3, and 3.35 GHz	Coupled Resonators	3	Many resonators are used

(Continued)

Table 1.1 Summary of multiband filters. (*Continued*)

Sl. no.	Author	Year	Resonant frequencies	Technique used	No. of bands	Limitations
10	Zhang et al. [22]	2007	1.84 and 2.9 GHz	Stub-Loaded Resonators	2	Minimum return loss
11	Zhao et al. [23]	2014	UWB with three notch band	E-shaped MMR (EMMR)	3	Insertion loss is not good
12	Pal et al. [24]	2014	3.5 GHz, 5.5 GHz, and 6.8 GHz	asymmetrically positioned SLOR	3	The roll-off rate is poor
13	Tsai et al. [25]	2014	2.4, 3.5, and 5.2 GHz	SIR	3	Insertion loss is not good
14	Hsu et al. [26]	2015	1.5, 2.5, 3.5 GHz	MMR	3	many resonators are used
15	Prashant et al. [68]	2018	2.85, 5.9 and 8.15 GHz	Triple-band stub-loaded open-loop resonator (TBSLOR)	3	Small bandwidth at lower frequency
16	P. Ranjan [69]	2019	2.4, 4.85, 7.93, and 9.75 GHz	Ω-shaped stub	4	Insertion loss is not good

Xu *et al.* [28] presented a UWB bandpass filter with Koch island-shaped stepped impedance lines (SIL). A Koch fractal–shaped ring slot is cut in the ground plane of the filter to realize negative permittivity. The composite right/left-handed transmission line is arranged with the gap in the conductor strip to realize negative permeability. It included five-section SIL on each side with an asymmetrical structure. The passband frequency range is 2.5–11 GHz with a relative bandwidth of 126%. Deng *et al.* [29] proposed a quintuple–mode stub-loaded resonator-based ultra-wideband bandpass filter. Two odd modes and three even modes are generated. Stepped impedance open and short stub are used to adjust the even–mode resonance frequencies but the odd–modes are fixed. Two TZs near the lower and upper cutoff frequencies can be generated by the short stub. High resonant modes of the desired passband are adjusted by applying a low–impedance line of the MMR. The open stubs are used to improve the upper stopband transmission zero. The passband frequency range of the filter is 2.8–11.2 GHz. Group delay variation and insertion loss are achieved lower than 2 dB and less than 0.63 ns, respectively.

Hao *et al.* [30] proposed a UWB filter based on multilayer technology. A transmission zero at the upper stopband has been generated by designing a resonator on the middle layer. Lower stopband transmission zero has been generated by designing a shorted coupled line on the top layer. Multilayer liquid crystal polymer technology is used to fabricate the filter. Bandwidth from 3 to 9 GHz is achieved with a flat group delay. It is useful for wireless UWB systems. Chu *et al.* [31] proposed a UWB bandpass filter based on stub–loaded MMR. The MMR is loaded with three open stubs. A stepped–impedance stub is positioned at the center and two stubs at the symmetrical side are located. Three even modes, two odd modes, and two transmission zeros are generated by the stepped–impedance stub. The resonator is designed to locate the two odd modes within the UWB band. Parameters of the stepped–impedance stub at the center can be used to adjust the even modes only. Passband frequency is 3.1–11.1 GHz, with a fractional bandwidth of 117%. Group delay within the UWB passband is between 0.25–0.70 ns.

Zhang *et al.* [32] proposed a UWB filter by using shorted stepped impedance stubs cascaded with the interdigital coupled line. Four even modes and three odd modes are generated. Odd and even mode analysis is used to verify the circuit. UWB filter covers passband from 3.4 to 10.7 GHz. Zhu *et al.* [33] presented a UWB bandpass filter based on dual-stub-loaded resonators (DSLR). Two transmission zeros are generated at the lower and upper stopband by applying The DSLR. Lengths of the stubs can be used to control the bandwidth. The relative dielectric constant of substrate material used to fabricate is 2.55 with a loss tangent of 0.0019 and the height of the substrate is 0.8 mm. Fractional BW of the filter is 106% and insertion loss is less than 0.7 dB. Li *et al.* [34] proposed two UWB bandpass filters based on an improved model. This paper designed two UWB BPFs with a fractional bandwidth of 51% (3.1 to 5.2 GHz) and 108% (3 to 10 GHz). Four short-circuited stubs connected with transmission lines are included to improve the model. The first and fourth short-circuited stubs are used to generate two transmission zeros. Less than 3 dB insertion loss is found between 3.1 to 10 GHz and less than 0.4 dB is found between 3.1 to 5.2 GHz. The sizes of the filters are 19 mm × 14 mm and 15 mm × 15 mm, respectively. Matrix analysis and short-circuited stubs model with the improved distributed quarter wave is presented.

Saadi *et al.* [35] proposed a design technique to implement UWB bandpass filters based on an integrated passive device (IPD) technology. Hourglass filter theory and the inductors are included in the filter circuit with the zigzag method for miniaturization of the filter. The filter is designed in 0.18 μm CMOS technology. This filter exhibits enhanced selectivity and controllable transmission zeros. Design approaches to this filter can be divided into two groups. The first one has established an electrical circuit model and the second is manufacturing technologies. The electrical circuit model is used to obtain the filter electrical specifications that affect the filter's physical aspects. Bandwidth covered the entire UWB spectrum. Taibi *et al.* [36] proposed, stepped–impedance open stub (SIS) based ultra-wideband bandpass filter. SIS is connected in the center of a uniform impedance transmission line. For coupling enhancement three interdigital parallel coupled-lines below aperture–backed are connected at each side of the filter. The frequency range of the filter is 3.2–11.1 GHz having a fractional bandwidth of 115%.

Janković *et al.* [37] proposed a defective ground structure-based UWB bandpass filter, which is used to connect a square patch with a ground plane to generate a resonant mode at a frequency lower than the without grounded patch resonator of two fundamental modes. Fundamental resonant frequencies can be controlled independently. By creating slots in the patch the higher modes resonant frequencies can be decreased. The Taconic CER10 substrate material is used with a relative dielectric constant of 9.8 and a thickness of 1.27mm. Group delay is 0.25 ns and insertion loss is less than 0.9 dB. Passband frequency for UWB band is 3.09–10.69 GHz. Yun *et al.* [38] proposed a particle swarm optimization (PSO) process to design a UWB bandpass filter. One cell CRLH–TL resonator, stepped impedance (SI), and two spur lines are used to design the UWB filter. The one-cell CRLH–TL resonator has a wide passband filtering characteristic.

The harmonics at the outside of the UWB band are removed by using one SI and two spur lines. Less than 1.4 dB flat insertion losses within passband and 11.5 to 22 GHz stopbands are achieved.

Sekaret *et al.* [40] proposed a slow-wave CPW based notch band UWB filter. The filter provided improved skirt rejection and better stopband rejection by DGS. DGS is used to achieve attenuation of the signal from 11 to 16 GHz. A notch is created by using a bridge structure to reject WLAN interference at 5.65 GHz. A summary of previous work on ultra-wideband filters is given in the following Table 1.2.

1.2.3   Ultra-Wideband Filter with Notch Band Characteristic

Rabbi *et al.* proposed [41] a UWB bandpass filter with a reconfigurable notched band to reject unwanted signals. A PIN diode is used as a switch for the notch. The filter has a notched band at 3.5 GHz when the switch is in the ON state. In the OFF state, a full band response is obtained. A third-order BPF consists of a single $\lambda_g/2$ resonator placed between two $\lambda_g/4$ short-circuited resonators. A grounded end is added with L–shaped parallel coupled transmission line to remove the undesired signal at 3.5 GHz. Zhao *et al.* [42] proposed a UWB bandpass filter using E–shaped resonator with two sharp notches. Genetic algorithm (GA) based UWB BPF is designed in which a set of structures as a chromosome and a structure as a gene is defined. The dual notch bands are generated by adjusting resonant frequencies of the E–shaped resonator. Two notches at 5.9 GHz and 8.0 GHz are obtained.

Song *et al.* [43] proposed a notched bands ultra-wideband bandpass filter based on triangular-shaped DGS. Transmission zeros are produced at a higher frequency by assigning six tapered defected ground structures. The low impedance microstrip is connected with short-circuited stubs to generate a TZ at the lower cut-off frequency. By increasing and folding the arm of the coupled-line to create a notch at 5.3 GHz, and create another notched band at 7.8 GHz by using a slot of the meander line. To achieve stronger coupling used quasi-IDC with slots on the ground plane. Sarkar *et al.* [44] proposed a UWB bandpass filter having high selectivity and dual notch bands. Short SLR and open stub are used to realize the UWB BPF. Meandered shorted stub is applied to the size of the filter. Two odd modes and two even modes excitations are present. Both modes are combined to achieve UWB BPF. Open stub loaded resonators (OSLR) are used to control the even mode frequencies and two transmission zeroes but odd mode frequencies are fixed. SLR is used to control the odd mode frequencies. Spiral resonators shaped slots of half-wavelength long are cut in the ground plane to obtained a notch band at 5.13 GHz. A notch at 8.0 GHz is

Table 1.2 Summary of ultra-wideband filters.

Sl. no.	Author	Year	Technique used	Band covered	No. of the modes present	Limitation
1	Sai *et al.* [27]	2009	MMR with stubs	2.8–11.0 GHz	4	Poor rejection skirt at upper cut off
2	Xu *et al.* [28]	2010	Koch island-shaped SI-lines (fractal)	2.5-11 GHz	2	Poor roll off rate
3	Hong *et al.* [29]	2010	Quintuple-mode stub-loaded resonator	2.8–11.2 GHz	5	Insertion loss is maximum
4	Hao *et al.* [30]	2011	multilayer liquid crystal polymer technology	3.0 to 9.05 GHz	2	The entire UWB band not covered
5	Qing *et al.* [31]	2011	Stub loaded MMR	3.1–11.1 GHz	5	Insertion loss is maximum
6	Zhang *et al.* [32]	2012	MMR with SI-stub	3.4 to 10.7 GHz	3	The entire UWB band not covered
7	He *et al.* [33]	2013	Dual-Stub-Loaded Resonator (DSLR)	2.9 to 10.9 GHz	3	Poor selectivity
8	Li *et al.* [34]	2014	Short-Circuited Stubs	3 to 10 GHz	4	The entire UWB band not covered

(Continued)

Table 1.2 Summary of ultra-wideband filters. (*Continued*)

Sl. no.	Author	Year	Technique used	Band covered	No. of the modes present	Limitation
9	Saadi *et al.* [35]	2015	Zigzag technique	3.1–10.6 GHz	2	Insertion loss is maximum
10	Taibi *et al.* [36]	2015	Stepped-impedance open stub (SIS)	3.1–10.6 GHz	3	Poor selectivity
11	Janković *et al.* [37]	2016	Grounded square patch resonator	3.09 to 10.69 GHz	3	Poor selectivity
12	Young *et al.* [38]	2016	Composite right- and left-handed-transmission line (CRLH-TL) resonator	3.1–10.6 GHz With notch	2	The poor roll-off rate at lower cut off
13	Wen *et al.* [39]	2010	DGS	3.1- 10.6 GHz	3	Notch band present
14	Vikram *et al.* [40]	2011	Slow-wave CPW MMR	3.1 to 10.6 GHz	3	The poor roll-off rate at lower cut off

achieved by adding an inward folded resonator near open SLR. Passband insertion loss is within 1.5dB. The notch bands eliminate the interference of WLAN and satellite frequency signal.

Chen *et al.* [45] proposed a narrow notch band UWB bandpass filter with sharp rejection. MMR with a stepped-impedance stub and a ring resonator is used to design the UWB BPF. The MMR has three transmission zeros (TZs) and six resonance frequencies. Stepped-impedance admittance ratios can be tuned to obtain notched bands. Bandwidth from 3.06 to 10.3 GHz and a notched band from 5.6 to 5.8 GHz is achieved. Mirzaee *et al.* [46] proposed a technique to control the notch band bandwidth in a UWB BPF. The technique is based on signal superposition. Folded trisection SIR is used between the input-output lines with electromagnetically coupled to active the signal transmission. Due to this technique, the bandwidth of the notch band can be increased by ∼370%. The filter exhibits a fractional bandwidth of 119.5% (2.96 to 11.75 GHz). Two notch bands are created at 5.2 and 8.9 GHz.

Yang *et al.* [47] proposed a DGS-based UWB BPF with the notched band. Cascading the IDC combined with H–shaped slots to design UWB BPF. The H–shaped slot is used to tighten the coupling of IDC. Six DGS slots are created to obtain their TZs towards the out of the UWB band. Low impedance microstrip is connected with short-circuited stub to generate a TZ at the lower cut off frequency. Combining H-shaped slots and six DGS slots to achieved circuit miniaturization. Undesired radio signals are rejected by adding a meander line slot.Less than 1.0 dB Insertion loss is found throughout the passband of 2.8 to 10.8 GHz. The center frequency of the notch band is 5.47 GHz. A wide stopband up to at least 20 GHz with 20 dB attenuation is achieved.

Chun *et al.* [48] proposed a reconfigurable UWB BPF with a switchable notch band. The filter consists of five stubs with short-circuited and four connecting lines on microstrip. Two identical switchable PIN diodes for electronic switching are used. When PIN diode is ON notch band is off and when PIN diode is off condition notch band is created with a center frequency of 5.1 GHz. Luo *et al.* [49] proposed a UWB BPF with a notched band by using a hybrid microstrip structure and CPW structure. Split resonant frequencies at the lower end, middle, and higher end of the UWB passband can be allocated by a detached–mode resonator (DMR). The CPW DMR consists of a short-stub loaded CPW resonator and a single-mode CPW resonator (SMCR). DMR is added with the meander slot-line structure to achieve the notch band and avoid interferences. The notched band is crated at 5.80 GHz.

Li *et al.* [50] proposed the topology of a notched band planar microstrip filter. A bandpass filter is used to fold short-circuited stubs to achieve better selectivity and good out-of-band characteristics. A narrow notched band is created by adding a defected SRR to reject the undesired WLAN radio signals.The center frequency of the notched band is 5.4 GHz. Song *et al.* [51] proposed an asymmetrically coupled UWB BPF with a multiple-notched-band. The dual-mode Y-shaped resonator has been used to design the UWB BPF. Multiple notched bands are created by asymmetrically coupled dual-line structures. Three passband bandwidths are 3–4.7 GHz, 5–8.1 GH), and 9–11.2 GH). The two notches are obtained at 4.9 and 8.55 GHz.

Kim *et al.* [52] proposed a ring resonator-based notched band UWB BPF. Two SIS and ring resonators are used to design the filter. Using SIS, improved return losses in a high–frequency band is obtained. The asymmetric structure of interdigital-coupled feed lines connected with the stepped-impedance ports is developed a notch band at 5 GHz. Luo *et al.* [53] proposed UWB bandpass filters having two and four notched bands. Broadside-coupled structure and CPW structure are used to design the filter. CPW resonator and meander line are connected in the DMR on the CPW layer to achieve a dual-notch band at 5.2 and 5.8 GHz simultaneously to cancel the WLAN signals. Meander defected microstrip structure is used to further created a two-notch band at 3.5 and 6.8 GHz to avoid the interferences for WiMAX and RF identification communication.

Ghatak *et al.* [54] proposed a band-notch UWB bandpass filter. A low-pass filter with SIR is cascaded by a high-pass filter to achieve a band notch characteristic.The notch frequency can be adjusted by adjusting the impedance ratio of the SIR. A band notch at 5.22 GHz is obtained. Less than 0.5 dB insertion loss is achieved within the passband 2.9–11.6 GHz.

Liu *et al.* [55] proposed a dual notch band UWB bandpass filter. Filter design is based on two mushroom–type electromagnetic bandgap (EBG) structures. The filter consists of two separated horn-shaped symmetric coupled lines. The broadside coupling between the coupled lines and the open-ended CPW on the ground have been used.Dimensions of the EBG structures can be adjusted to obtain the notch bands frequencies at 5.2 and 5.8 GHz. Wu *et al.* [56] presented the implementation of UWB filters with tunable notch band-based integrated passive device technology. The filter is designed on a micromachined silicon substrate. The filter size is only 2.9 mm × 2.4 mm.

Wang *et al.* [57] proposed a cross-shaped UWB bandpass filter having a notched band. Quarter–wavelength short-circuited shunt stubs loaded with a cross-shaped resonator are used to realize the filter. Using open stubs a notch band is created at 8.0 GHz. Step–impedance open-circuited lines are used to realize the sharp skirt. The FBW of 119.4% (2.7–10.7 GHz)

Table 1.3 Summary of ultra-wideband filters with notched band.

Sl. no.	Author	Year	Notch band frequencies (in GHz)	No. of notch band	Technique used	Limitation
1	Lin *et al.* [39]	2010	5.5 GHz	1	DGS and the open stub	Poor roll-off rate
2	Wei *et al.* [62]	2010	5.75 and 8.05 GHz	2	DGS and SIS	Poor selectivity
3	Suyang *et al.* [63]	2012	2.5 GHz	1	slot-line DGS	Poor insertion loss
4	Zhang *et al.* [64]	2013	5.8 GHz	1	MMR with shorting stub	Poor selectivity
5	Chen *et al.* [65]	2014	5.7 GHz	1	MMR with stub	The entire UWB band not covered
6	Sandip *et al.* [66]	2016	3.6, 5.2, and 8.4 GHz	3	C-Shaped and E-Shaped Resonator	Poor insertion loss
7	A Kamma *et al.* [67]	2017	4.42, 5.54, 7.64 GHz	3	MMR with shorting stub	Via is used
8	Prashant *et al.* [70]	2017	5.9 GHz	1	Inverted T-shapedMMR	Insertion loss at second pass band is not good

is achieved. Xu *et al.* [58] proposed a notched band UWB filter by using RSLR (radial stub loaded resonator). The RSLR can be used to provide a notched band and control transmission zeros separately. UWB BPFs with a notched band at 8.0 GHz and without notch band BPF are presented. Multiple-mode resonant characteristic is excited on both sides of the RSLR by the coupled–line to achieve UWB performance. The Y–shaped RSLR is used to produce a TZ within the frequency range of 3.1 to 10.6GHz.

Xu *et al.* [59] proposed a UWB bandpass filter structure with a notch band. The design structure involved two shunts coupled-line on both sides and two coupled–line series in the mid-point. TZs on both sides of the passband can be generated by using shunt shorted coupled–lines. A notch band is produced at 8.0GHz by using an open stub for rejecting satellite communication signals. Sarkar *et al.* [60] proposed a dual notch band UWB bandpass filter. A modified distributed highpass filter (HPF) with defected stepped impedance resonator is used to suppress higher-order harmonics of the HPF and achieve UWB bandpass characteristic. By adding stubs in the coupled line two-notch band dual at 5.75 GHz and 8.05 GHz areobtained. Wei *et al.* [61] proposed a compact notch band UWB BPF by using a triple–mode SIR. The filter consists of two SIRs and two short-circuited stubs on its center plane. Triple–mode SIR is responsible for producing notched bands and providing a degree of freedom to control the resonant frequencies. Stopband attenuation up to 20 GHz and passband from 2.8–11.0 GHz is achieved. A summary of previous work on ultra-wideband filters with the notched band is given in the following Table 1.3.

1.3   Conclusions

The review of literature in this chapter has concentrated largely on empirical observations of filter design for UWB applications. Various design techniques reported in the literature for the multiband filter, UWB filter, UWB filter with notch band, and their design issues have been reviewed and discussed. Various authors have explained insertion loss minimization, transmission zeros creation, selectivity improvement, multiple band creation, notch band creation, bandwidth enhancement technique based on stub loaded open-loop resonator, split ring resonator, defected ground structure have been investigated in detail. It is found that sharp selectivity, bandwidth enhancement, size reduction, and interference avoidance are important issues related to the UWB filter.

References

1. Zuhair M. Hejazi and Zeshan Ali, "Multiband Bandpass Filters with Suppressed Harmonics Using a Novel Defected Ground Structure," *Microwave and Optical Technology Letters*, vol. 56, no. 11, November 2014.
2. C. F. Chen, T. Y. Huang and R. B. Wu, "Design of Dual- and Triple-Passband Filters Using Alternately Cascaded Multiband Resonators," in *IEEE Transactions on Microwave Theory and Techniques*, vol. 54, no. 9, pp. 3550-3558, Sept. 2006.
3. H. Di, B. Wu, X. Lai and C. H. Liang, "Synthesis of Cross-Coupled Triple-Passband Filters Based on Frequency Transformation," in *IEEE Microwave and Wireless Components Letters*, vol. 20, no. 8, pp. 432-434, Aug. 2010.
4. Ko-Wen Hsu, Wen-Hua Tu, "Design of asymmetrical resonator for microstrip triple-band and broadband bandpass filters" *Microelectronics Journal*, vol. 46, 2015.
5. C. Y. Liou and S. G. Mao, "Triple-Band MarchandBalun Filter Using Coupled-Line Admmittance Inverter Technique," in *IEEE Transactions on Microwave Theory and Techniques*, vol. 61, no. 11, pp. 3846-3852, November 2013.
6. J. Ai, Y. Zhang, K. D. Xu, Y. Guo, W. T. Joines and Q. H. Liu, "Compact sext-band bandpass filter based on single multimode resonator with high band-to-band isolations," in *Electronics Letters*, vol. 52, no. 9, pp. 729-731, April 2016.
7. Jia-Sheng Hong and M. J. Lancaster, "Couplings of microstrip square open-loop resonators for cross-coupled planar microwave filters," in *IEEE Transactions on Microwave Theory and Techniques*, vol. 44, no. 11, pp. 2099-2109, Nov. 1996.
8. Jen-Tsai Kuo, Tsung-HsunYeh and Chun-Cheng Yeh, "Design of microstrip bandpass filters with a dual-passband response," in *IEEE Transactions on Microwave Theory and Techniques*, vol. 53, no. 4, pp. 1331-1337, April 2005.
9. Xuehui Guan, Zhewang Ma and PengCai, "A novel triple-band microstrip Bandpass filter for wireless Communication," *Microwave and Optical Technology Letters*, vol. 51, no. 6, June 2009.
10. Wen Ko and Man-Long Her, "A Compact Coupled-Line Triple-Band Bandpass Filter with an Adjustable Third Passband and Open Stubs," *Microwave and Optical Technology Letters*, vol. 55, no. 5, pp. 970-975, May 2013.
11. X. M. Lin and Q. X. Chu, "Design of Triple-band Bandpass Filter Using Tri-section Stepped-Impedance Resonators," *2007 International Conference on Microwave and Millimeter Wave Technology, Builin*, 2007, pp. 1-3.
12. G. Wibisono and T. Syafraditya, "Triple band bandpass filter with cascade tri section stepped impedance resonator," *2014 1st International Conference on Information Technology, Computer, and Electrical Engineering, Semarang*, 2014, pp. 111-114.

13. R. H. Geschke, B. Jokanovic and P. Meyer, "Filter Parameter Extraction for Triple-Band Composite Split-Ring Resonators and Filters," in *IEEE Transactions on Microwave Theory and Techniques*, vol. 59, no. 6, pp. 1500-1508, June 2011.

14. H. Liu, J. Lei, X. Guan, L. Sun and Y. He, "Compact Triple-Band High-Temperature Superconducting Filter Using Multimode Stub-Loaded Resonator for ISM, WiMAX, and WLAN Applications," in *IEEE Transactions on Applied Superconductivity*, vol. 23, no. 6, pp. 99-103, Dec. 2013.

15. Q. Liu, Y. Liu, J. Shen, S. Li, C. Yu and Y. Lu, "Wideband Single-Layer 90° Phase Shifter Using Stepped Impedance Open Stub and Coupled-Line With Weak Coupling," in *IEEE Microwave and Wireless Components Letters*, vol. 24, no. 3, pp. 176-178, March 2014.

16. H. Liu, Y. Peng, B. Ren, P. Wen, J. Lei and X. Guan, "Compact Triple-Band Superconducting Filter Based on a Multimode Stepped-Impedance Split-Ring Resonator," in *IEEE Transactions on Applied Superconductivity*, vol. 26, no. 6, pp. 1-5, Sept. 2016.

17. XuemeiZheng and Tao Jiang, "A novel triple notch-bands UWB band-pass filter using E-shaped resonator and open-load stubs," 2015 *IEEE 6th International Symposium on Microwave, Antenna, Propagation, and EMC Technologies (MAPE), Shanghai*, 2015, pp. 570-573.

18. X. Guan et al., "Compact Triple-Band High-Temperature Superconducting Filter Using Coupled-Line Stepped Impedance Resonator," in *IEEE Transactions on Applied Superconductivity*, vol. 26, no. 7, pp. 1-5, Oct. 2016.

19. C. F. Chen, "Design of a Compact Microstrip Quint-Band Filter Based on the Tri-Mode Stub-Loaded Stepped-Impedance Resonators," in *IEEE Microwave and Wireless Components Letters*, vol. 22, no. 7, pp. 357-359, July 2012.

20. W. H. Tu and K. W. Hsu, "Design of Sext-Band Bandpass Filter and Sextaplexer Using Semilumped Resonators for System in a Package," in *IEEE Transactions on Components, Packaging and Manufacturing Technology*, vol. 5, no. 2, pp. 265-273, Feb. 2015.

21. M. Mokhtaari, J. Bornemann, K. Rambabu and S. Amari, "Coupling-Matrix Design of Dual and Triple Passband Filters," in *IEEE Transactions on Microwave Theory and Techniques*, vol. 54, no. 11, pp. 3940-3946, Nov. 2006.

22. X. Y. Zhang, J. X. Chen, Q. Xue and S. M. Li, "Dual-Band Bandpass Filters Using Stub-Loaded Resonators," in *IEEE Microwave and Wireless Components Letters*, vol. 17, no. 8, pp. 583-585, Aug. 2007.

23. J. Wang, J. Zhao and J. L. Li, "Compact UWB bandpass filter with triple notched bands using parallel U-shaped defected microstrip structure," in *Electronics Letters*, vol. 50, no. 2, pp. 89-91, January 16, 2014.

24. M. Pal, Joysmita Chatterjee, RowdraGhatak "Multiband BPF with wide upper stop band using asymmetrically positioned stub loaded open loop resonators" *Int. J. Electron. Commun.* (AEÜ) 68, 1041–1046, 2014.

25. Lin-Chuan Tsai, "Design of compact triple passband microwave filters based on stepped impedance resonators" *Microwave and Optical Technology Letters*, Vol. 57, No. 1, January 2015.

26. Ko-Wen Hsu, Wen-Hua Tu, "Design of asymmetrical resonator for microstrip triple-band and broad band bandpass filters", *Microelectron. J*, 2015.

27. S. W. Wong and L. Zhu, "Quadruple-Mode UWB Bandpass Filter With Improved Out-of-Band Rejection," in *IEEE Microwave and Wireless Components Letters*, vol. 19, no. 3, pp. 152-154, March 2009.

28. H. X. Xu, G. M. Wang and C. X. Zhang, "Fractal-shaped UWB bandpass filter based on composite right/left handed transmission line," in *Electronics Letters*, vol. 46, no. 4, pp. 285-287, February 18, 2010.

29. H. w. Deng, Y. j. Zhao, L. Zhang, X. s. Zhang and S. p. Gao, "Compact Quintuple-Mode Stub-Loaded Resonator and UWB Filter," in *IEEE Microwave and Wireless Components Letters*, vol. 20, no. 8, pp. 438-440, Aug. 2010.

30. Z. C. Hao and J. S. Hong, "High selectivity UWB bandpass filter using dual-mode resonators," in *Electronics Letters*, vol. 47, no. 25, pp. 1379-1381, December 8, 2011.

31. Q. X. Chu, X. H. Wu and X. K. Tian, "Novel UWB Bandpass Filter Using Stub-Loaded Multiple-Mode Resonator," in *IEEE Microwave and Wireless Components Letters*, vol. 21, no. 8, pp. 403-405, Aug. 2011.

32. Z. Zhang and F. Xiao, "An UWB Bandpass Filter Based on a Novel Type of Multi-Mode Resonator," in *IEEE Microwave and Wireless Components Letters*, vol. 22, no. 10, pp. 506-508, Oct. 2012.

33. H. Zhu and Q. X. Chu, "Compact Ultra-Wideband (UWB) Bandpass Filter Using Dual-Stub-Loaded Resonator (DSLR)," in *IEEE Microwave and Wireless Components Letters*, vol. 23, no. 10, pp. 527-529, Oct. 2013.

34. X. Li and X. Ji, "Novel Compact UWB Bandpass Filters Design With Cross-Coupling Between λ/4 Short-Circuited Stubs," in *IEEE Microwave and Wireless Components Letters*, vol. 24, no. 1, pp. 23-25, Jan. 2014.

35. A. A. Saadi, M. C. E. Yagoub, R. Touhami, A. Slimane, A. Taibi and M. T. Belaroussi, "Efficient UWB filter design technique for integrated passive device implementation," in *Electronics Letters*, vol. 51, no. 14, pp. 1087-1089, Jul. 9, 2015.

36. A. Taibi, M. Trabelsi, A. Slimane, M. T. Belaroussi and J. P. Raskin, "A Novel Design Method for Compact UWB Bandpass Filters," in *IEEE Microwave and Wireless Components Letters*, vol. 25, no. 1, pp. 4-6, Jan. 2015.

37. N. Janković, G. Niarchos and V. Crnojević-Bengin, "Compact UWB band-pass filter based on grounded square patch resonator," in *Electronics Letters*, vol. 52, no. 5, pp. 372-374, 3 3 2016.

38. Y. C. Yun, S. H. Oh, J. H. Lee, K. Choi, T. K. Chung and H. S. Kim, "Optimal Design of a Compact Filter for UWB Applications Using an Improved Particle Swarm Optimization," in *IEEE Transactions on Magnetics*, vol. 52, no. 3, pp. 1-4, March 2016.

39. W. J. Lin, J. Y. Li, L. S. Chen, D. B. Lin and M. P. Houng, "Investigation in Open Circuited Metal Lines Embedded in Defected Ground Structure and Its Applications to UWB Filters," in *IEEE Microwave and Wireless Components Letters*, vol. 20, no. 3, pp. 148-150, March 2010.
40. V. Sekar and K. Entesari, "Miniaturized UWB Bandpass Filters With Notch Using Slow-Wave CPW Multiple-Mode Resonators," in *IEEE Microwave and Wireless Components Letters*, vol. 21, no. 2, pp. 80-82, Feb. 2011.
41. K. Rabbi, L. Athukorala, C. Panagamuwa, J. C. Vardaxoglou and D. Budimir, "Compact UWB bandpass filter with reconfigurable notched band," in *Electronics Letters*, vol. 49, no. 11, pp. 709-711, May 23, 2013.
42. J. Zhao, J. Wang, G. Zhang and J. L. Li, "Compact Microstrip UWB Bandpass Filter With Dual Notched Bands Using E-Shaped Resonator," in *IEEE Microwave and Wireless Components Letters*, vol. 23, no. 12, pp. 638-640, Dec. 2013.
43. Y. Song, G. M. Yang and W. Geyi, "Compact UWB Bandpass Filter With Dual Notched Bands Using Defected Ground Structures," in *IEEE Microwave and Wireless Components Letters*, vol. 24, no. 4, pp. 230-232, April 2014.
44. P. Sarkar, R. Ghatak, M. Pal and D. R. Poddar, "High-Selective Compact UWB Bandpass Filter With Dual Notch Bands," in *IEEE Microwave and Wireless Components Letters*, vol. 24, no. 7, pp. 448-450, July 2014.
45. X. Chen, L. Zhang and Y. Peng, "UWB bandpass filter with sharp rejection and narrow notched band," in *Electronics Letters*, vol. 50, no. 15, pp. 1077-1079, July 17, 2014.
46. M. Mirzaee, S. Noghanian and B. S. Virdee, "High selectivity UWB bandpass filter with controllable bandwidth of dual notch bands," in *Electronics Letters*, vol. 50, no. 19, pp. 1358-1359, September 11, 2014.
47. G. M. Yang, R. Jin, C. Vittoria, V. G. Harris and N. X. Sun, "Small Ultra-Wideband (UWB) Bandpass Filter With Notched Band," in *IEEE Microwave and Wireless Components Letters*, vol. 18, no. 3, pp. 176-178, March 2008.
48. Y. H. Chun, H. Shaman and J. S. Hong, "Switchable Embedded Notch Structure for UWB Bandpass Filter," in *IEEE Microwave and Wireless Components Letters*, vol. 18, no. 9, pp. 590-592, Sept. 2008.
49. X. Luo, H. Qian, J. G. Ma, K. Ma and K. S. Yeo, "A compact UWB bandpass filter with ultra narrow notched band and competitive attenuation slope," *2010 IEEE MTT-S International Microwave Symposium, Anaheim, CA*, 2010, pp. 221-224.
50. Q. Li, Z. J. Li, C. H. Liang and B. Wu, "UWB bandpass filter with notched band using DSRR," in *Electronics Letters*, vol. 46, no. 10, pp. 692-693, May 13, 2010.
51. K. Song and Q. Xue, "Asymmetric dual-line coupling structure for multiple-notch implementation in UWB bandpass filters," in *Electronics Letters*, vol. 46, no. 20, pp. 1388-1390, September 30, 2010.

52. C. H. Kim and K. Chang, "Ultra-Wideband (UWB) Ring Resonator Bandpass Filter With a Notched Band," in *IEEE Microwave and Wireless Components Letters,* vol. 21, no. 4, pp. 206-208, April 2011.

53. X. Luo, J. G. Ma, K. S. Yeo and E. P. Li, "Compact Ultra-Wideband (UWB) Bandpass Filter With Ultra-Narrow Dual- and Quad-Notched Bands," in *IEEE Transactions on Microwave Theory and Techniques*, vol. 59, no. 6, pp. 1509-1519, June 2011.

54. R. Ghatak, P. Sarkar, R. K. Mishra and D. R. Poddar, "A Compact UWB Bandpass Filter With Embedded SIR as Band Notch Structure," in *IEEE Microwave and Wireless Components Letters*, vol. 21, no. 5, pp. 261-263, May 2011.

55. B. W. Liu, Y. Z. Yin, Y. Yang, S. H. Jing and A. F. Sun, "Compact UWB bandpass filter with two notched bands based on electromagnetic bandgap structures," in *Electronics Letters*, vol. 47, no. 13, pp. 757-758, June 23, 2011.

56. Z. Wu, Y. Shim and M. Rais-Zadeh, "Miniaturized UWB Filters Integrated With Tunable Notch Filters Using a Silicon-Based Integrated Passive Device Technology," in *IEEE Transactions on Microwave Theory and Techniques*, vol. 60, no. 3, pp. 518-527, March 2012.

57. H. Wang, W. Kang, C. Miao and W. Wu, "Cross-shaped UWB bandpass filter with sharp skirt and notched band," in *Electronics Letters*, vol. 48, no. 2, pp. 96-97, January 19, 2012.

58. J. Xu, W. Wu, W. Kang and C. Miao, "Compact UWB Bandpass Filter With a Notched Band Using Radial Stub Loaded Resonator," in *IEEE Microwave and Wireless Components Letters*, vol. 22, no. 7, pp. 351-353, July 2012.

59. J. Xu, W. Kang, C. Miao and W. Wu, "Sharp rejection UWB bandpass filter with notched band," in *Electronics Letters*, vol. 48, no. 16, pp. 1005-1006, August 2, 2012.

60. P. Sarkar, R. Ghatak, M. Pal and D. R. Poddar, "Compact UWB Bandpass Filter With Dual Notch Bands Using Open Circuited Stubs," in *IEEE Microwave and Wireless Components Letters*, vol. 22, no. 9, pp. 453-455, Sept. 2012.

61. F. Wei, W. T. Li, X. W. Shi and Q. L. Huang, "Compact UWB Bandpass Filter With Triple-Notched Bands Using Triple-mode Stepped Impedance Resonator," in *IEEE Microwave and Wireless Components Letters*, vol. 22, no. 10, pp. 512-514, Oct. 2012.

62. F. Wei, C. J. Gao, B. Liu, H. W. Zhang and X. W. Shi, "UWB bandpass filter with two notch-bands based on SCRLH resonator," in *Electronics Letters*, vol. 46, no. 16, pp. 1134-1135, August 5, 2010.

63. S. Shi, W. W. Choi, W. Che, K. W. Tam and Q. Xue, "Ultra-Wideband Differential Bandpass Filter With Narrow Notched Band and Improved Common-Mode Suppression by DGS," in *IEEE Microwave and Wireless Components Letters*, vol. 22, no. 4, pp. 185-187, April 2012.

64. X. Y. Zhang, Y. W. Zhang and Q. Xue, "Compact Band-Notched UWB Filter Using Parallel Resonators With a Dielectric Overlay," in *IEEE Microwave and Wireless Components Letters*, vol. 23, no. 5, pp. 252-254, May 2013.

65. D. Cheng , H.-C. Yin & H.-X. Zheng (2012) A compact dual-band bandstop filter with defected microstrip slot," *Journal of Electromagnetic Waves and Applications*, 26:10, 1374-1380.

66. S. Kumar, R. D. Gupta and M. S. Parihar, "Multiple Band Notched Filter Using C-Shaped and E-Shaped Resonator for UWB Applications," in *IEEE Microwave and Wireless Components Letters*, vol. 26, no. 5, pp. 340-342, May 2016.

67. A. Kamma, R. Das, D. Bhatt and J. Mukherjee, "Multi Mode Resonators Based Triple Band Notch UWB Filter," in *IEEE Microwave and Wireless Components Letters*, vol. 27, no. 2, pp. 120-122, Feb. 2017.

68. Prashant Ranjan,V. S. Tripathi, "A Compact Triple Band Microwave Filter Using Symmetrically Placed Stub Loaded Open-loop Resonators," *International Journal of Electronics Letter* (Taylor & Francis), Vol. 6, No. 3, pp. 364–375, 2018.

69. Prashant Ranjan, "Triple-band and multiband filters using SIL and MSLOR for wireless communication system," *International Journal of Electronics Letter* (Taylor & Francis), pp. 1-15, 2019.

70. Prashant Ranjan, N. Kishore, V. K. Dwivedi, G. Upadhyay and V. S. Tripathi, "UWB filter with controllable notch band and higher stop band transmission zero using open stub in inverted T-shaped resonator," *2017 IEEE Asia Pacific Microwave Conference (APMC), Kuala Lumpur, Malaysia*, 2017, pp. 817-820.

Design, Isolation Analysis, and Characterization of 2×2/4×4 MIMO Antennas for High-Speed Wireless Applications

Manish Sharma[1]*, Rajeev Kumar[1] and Preet Kaur[2]

[1]*Chitkara University Institute of Engineering and Technology Chitkara University, Punjab, India*
[2]*JC Bose University of Science and Technology, YMCA, Faridabad, Haryana, India*

Abstract

A single antenna transmitting and receiving radio frequency signals (RF) suffers demerits such as multiple fading which not only reduces the bandwidth but also efficiency. Fading phenomenon impacts the signal-to-noise ratio (SNR) and hence also affects the error rate of the digital signal transmitted. The solution to the above problem can be found by using multiple radiating elements on the same patch which reduces the effects of multiple path fading and is known as Multiple-Input-Multiple-Output (MIMO) antenna system configuration. This MIMO antenna configuration works on different diversity techniques known as time, frequency, and spatial. Also, by using multiple antennas, the bandwidth of the MIMO antenna system is increased which is defined by Shannon's channel capacity. This chapter discussion is focused on 2×2 and 4×4 MIMO antenna configuration and reference antenna are taken which is characterized in both frequency and diversity domain. In designing of MIMO antenna configuration, one has to deal with the challenge of maintaining the operating bandwidth as well as isolation between the radiating elements. This chapter focuses on different techniques used and characterization of MIMO diversity performance in terms of Envelope Correlation Coefficient≤0.2, Diversity Gain 9.95dB, Total Active Reflection Coefficient <0dB, and Channel Capacity Los <0.2b/s/Hz.

**Corresponding author*: manishengineer1978@gmail.com

Prashant Ranjan, Dharmendra Kumar Jhariya, Manoj Gupta, Krishna Kumar, and Pradeep Kumar (eds.)
Next-Generation Antennas: Advances and Challenges, (23–48) © 2021 Scrivener Publishing LLC

Keywords: Multiple path fading, MIMO, diversity, Shannon's channel capacity, 2×2/4×4 MIMO, ECC, DG, TARC, CCL

2.1   Introduction

In today's communication systems, antennas play a major role, responsible for the transition of electromagnetic waves and converting it to induce an alternating current in the transmission environment (Transmitter section) or converting induced current to electromagnetic waves (Receiver section). However, several Single-Input-Single-Output (SISO) with one radiating element have been proposed for communication system but this suffers drawbacks due to multiple path fading, which not only decreases the operating bandwidth of interest but also reduces the efficiency. Using multiple radiating elements on the same patch forming Multiple-Input-Multiple-Output (MIMO) configuration provides the solution for demerits of the SISO system. However, designing this MIMO antenna configuration is a challenging task in planar technology as different aspects of the design have to be encountered. Ultrawideband (UWB) MIMO antennas [1–11] with 2×2 configuration show different isolation techniques used to obtain better transmission coefficient (S_{12}, S_{21}, S_{31}, S_{32}, S_{34}, S_{31}, S_{32}, S_{34}, S_{41}, S_{42}, S_{43}, S_{14}, S_{24}, S_{34} for 4×4 MIMO) and thus different isolation techniques better than 20dB are reported. 5G (5th generation) MIMO antenna operating at 7GHz [1] provides isolation of more than 20dB where the rod-like parasitic element is added backed plane. A slotted rectangular patch [2] added to the ground, rectangular strip etched in-ground and placed 45° orientation [3] and L-T type stub in-ground [4–6] also provides better isolation in designing of MIMO antenna. Etching two symmetrical slots on the ground plane [7], slotted T-type stub in-ground [8–10], and placing radiating elements adjacent/orthogonally with separated L-type slots in the ground [11] achieves better isolation between two ports of UWB-MIMO antenna. MIMO antenna with 3 cell configuration of compact size 0.48λo×1.34 λo×0.05λo ensures isolation less than -25dB [12]. Loading of H-type DGS (defected ground structure) provides isolation more than 20dB [20] and in pattern diversity MIMO antenna, four L-shaped branches attached to circular patch placed counter-clockwise also provides good isolation between the bandwidth of 4.85GHz-6.12GHz. Dual polarization MIMO antenna [15–20] sharing the common patch with two feedings and the respective ground plane not only results in dual-polarization but the etched slot on patch and stub attached to ground results in good isolation. Also, using Electromagnetic Band Gap (EBG) structure and using Dielectric Resonator Antennas (DRAs) in MIMO antenna configuration are other reported techniques for isolation [21, 22]. In Ultrawideband, single notched MIMO antenna [24–34] with different isolation methods have been used to achieve higher isolation which includes rectangular stub in the ground with 45° orientation [24], T-type stub, and L-type in-ground [26]. Also, dual notched MIMO antenna [35–43] intended for UWB applications uses different techniques for isolation such as slotted ground [35], funnel-type slot [39], and T-type stub in-ground [43]. Also, the MIMO antenna capable of rejecting three interfering bands [44–47] and MIMO antenna with four notched bands [48] records better isolation between radiating patch-elements by using slotted ground and T-type stub [49]. 4×4 MIMO antenna configuration [50–73] which are segregated as UWB, UWB-single notch, UWB-dual notch, and UWB-triple notch utilizes different isolation techniques such as placing of radiating elements orthogonally, using neutralization technique as stub connected between radiating patch, fan-type parasitic decoupling elements are also isolation techniques used. 8-port and 10-port MIMO antenna [74, 75] also successfully achieves good isolation.

This chapter focuses on the study of MIMO antenna configuration with two and four radiating elements forming 2×2 and 4×4 antenna configuration. Also, different techniques to achieve higher isolation are discussed and a comparison is made in terms of physical size and diversity performance for both the configurations.

2.2   Understanding 2×2 MIMO Antenna Configuration

Developing of MIMO antenna configuration is carried out by designing a single unit MIMO configuration (1×1 cell) and then converting the design to the desired M×N (M=1,2,3…., N=1,2,3…) MIMO configuration as per the requirement and applications. To understand the above concept with characterization and Diversity performance, a 2×2 MIMO antenna [51] is taken as a reference. Initially, as observed from Figure 2.1(a) showing rotated view of a single element antenna signifies a single radiating patch with the ground. The radiating patch is connected to a 50Ω microstrip transmission line and is fed by a matched SMA connector for input RF signals. Taconic RF-30 Microwave substrate is used in designing the antenna with electrical properties including permittivity εr=3.0 and loss tangent of 0.0014 with height h=1.52mm. The modified square patch is printed on one plane and

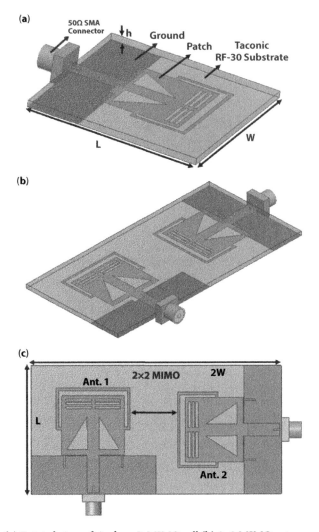

Figure 2.1 (a) Rotated view of single unit MIMO cell (b) 2×2 MIMO antenna rotated view (c) Arrangement of radiating elements with respective ground.

ground on the opposite providing operational bandwidth of 2.85GHz-11.52GHz. Also, a U-type stub added to the rectangular patch calculated at a guided wavelength of 25mm corresponding to 7GHz results in a lower cut-off frequency to achieve UWB bandwidth. SMA connector is shown in Figure 2.1(a) is actually modeled in an RF simulator to match the result with measured values. Figure 2.1(b) represents the modified version resulting in a 2×2 MIMO antenna configuration. As per the observation, two identical radiating elements are placed by a distance of K=15.2mm ensuring minimal spacing between them and in an orthogonal fashion (respective axis-aligned 90°). Modification in the rectangular patch (3 pair of etched rectangular slots and pair of the mirror-imaged right-angled triangular slot) with three slots etched in-ground is carried out to improve the matching of impedance and to obtain wider operating bandwidth. 2×2 MIMO antenna shown in Figure 2.1(c) has to verify for diversity performance and is discussed below.

Figure 2.2 characterizes the 2×2 MIMO antenna configuration shown in Figure 2.1(b)-(c). Figure 2.2(a) shows the plot of the real and imaginary

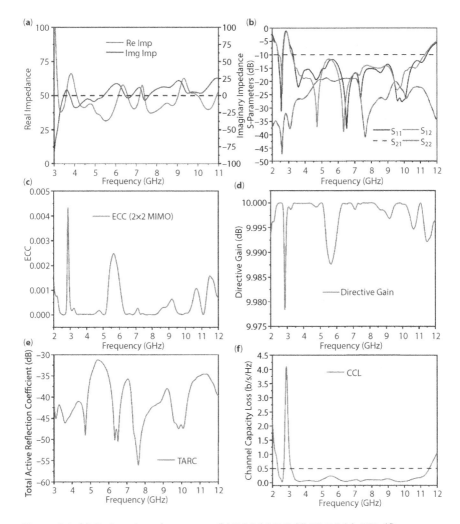

Figure 2.2 (a) Re-Img, impedance curve (b) ECC (c) DG (d) TARC (e) CCL (f).

impedance curve for the MIMO configuration. In an ideal condition, real impedance is matched with 50Ω and imaginary impedance value is 0Ω. Practically, this is not the case and there is a slight mismatch of impedance as observed from Figure 2.2(a). It can be approximated that the real and imaginary impedance values are close to the ideal values of impedance and hence a good impedance bandwidth is obtained for a wider frequency range. The diversity performance of the antenna is based on the reflection and transmission coefficients of the MIMO antenna configuration. Figure 2.2(b) shows S_{11}, S_{12}, S_{21}, and S_{22} for the 2×2 MIMO antenna configuration. As per the impedance, bandwidth is concerned, designed MIMO is well suited for UWB applications and also maintains isolation of more than 20dB.

This itself suggests that the MIMO antenna will show better diversity performance. In multiple channel scenario with the enriched signal, increases the capacity linearly of MIMO antenna configuration of elements says D in D×D MIMO configuration. It is also a known fact that in the receiver MIMO antenna system, the capacity is reduced practically due to the correlation between the signals received by the receiver from different angles. Hence, from the MIMO design aspect, correlation of the signals received from different radiating patches becomes more important and hence, Envelope Correlation Coefficient (ECC) has to be limited to 0.5. This value is achieved in the above-discussed MIMO antenna and Figure 2.2(c) suggests that the values are well below the permissible values for the entire operating band. Required ECC can not only be calculated irrespective of S-Parameters but also using far-field patterns. For 2×2 MIMO antenna, the far-field pattern is given by

$$ECC = \frac{\left| \iint\limits_{4\pi} \left[\vec{F_a}(\theta,\phi) \times \vec{F_b^*}(\theta,\phi) \right] d\Omega \right|^2}{\iint\limits_{4\pi} \left| \vec{F_a}(\theta,\phi) \right|^2 d\Omega \iint\limits_{4\pi} \left| \vec{F_b}(\theta,\phi) \right|^2 d\Omega} \tag{2.1}$$

The above equation signifies the calculation of ECC for two antenna configurations. Considering excitation of the ith port and assuming all the termination of the ports to matched 50Ω. Also, the operator • in the above equation signifies the Hermitian product. This equation is derived from the ECC between the two antennas. In general, ECC between any two ports can be calculated. The relation between S-parameter and the radiation pattern can be related as

$$\overrightarrow{I} - \overrightarrow{S}^{\overrightarrow{H}^{\star}} \overrightarrow{S} = \overrightarrow{F}^{\overrightarrow{H}^{\star}} \overrightarrow{F} \tag{2.2}$$

In the above equation, I is the identity matrix and $H^{\star}$ denotes the transpose of the complex conjugate. Let us assume MIMO configuration consisting of T antennas as shown in Figure 2.3. Then, ECC between two antenna elements, say M and N is calculated by

$$ECC(M,N,T) = \frac{\left|C_{m,n}(T)\right|^{2}}{\prod\limits_{P=M,N}[1-CP,P(N)]} \tag{2.3}$$

where

$$C_{M,N}(T) = \sum_{t=1}^{T} S_{M,t}^{\star} S_{t,N} \tag{2.4}$$

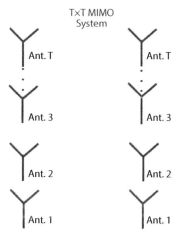

Figure 2.3 Arrangement of T×T MIMO antenna configuration.

From the scattering parameter, ECC is calculated by the following equation

$$ECC(M,N,T) = \frac{\left| \sum\limits_{t=1}^{T} S_{M,t}^{\star} S_{t,N} \right|^2}{\prod\limits_{P=M,N} \left[1 - \sum\limits_{t=1}^{T} S_{P,t}^{\star} S_{t,P} \right]} \tag{2.5}$$

Thus with 2×2 MIMO antenna configuration, ECC between Antenna 1 (M=1) and Antenna 2 (N=2) can be written as

$$ECC = \frac{\left| s_{11}^{\star} s_{12} + s_{21}^{\star} s_{22} \right|^2}{\left(\left(1 - |s_{11}|^2 - |s_{21}|^2\right)\left(1 - |s_{22}|^2 - |s_{12}|^2\right) \right)} \tag{2.6}$$

When MIMO configuration is actively working, due to placement of radiating elements close to each other (minimum distance λ/2) performance of the MIMO system can be affected due to interaction of radiating elements with each other. It is also known that the isolation values do define the degree of interferences but only these values are not sufficient to decide the efficient working of any MIO configuration. This leads to the requirement of defining a new metric called Total Active Reflection Coefficient (TARC) and is defined as "square root of the ratio of total reflected power to the total incident power and apparent loss of the overall MIMO antenna system". As per the definition, TARC of 2×2 MIMO configuration is calculated as

$$TARC_{2\times2} = \sqrt{\frac{(S_{11} + S_{12})^2 + (S_{21} + S_{22})^2}{2}} \tag{2.7}$$

Also for any MIMO configuration, TARC is always <0dB and is calculated fir -10dB bandwidth.

Diversity Gain (DG) is another important used for characterization of MIMO antenna and ideally should be >9.95dB. DG is a parameter that is dependent on ECC and is calculated by the following equation

$$DG_{2\times2} = 10\sqrt{1 - ECC^2} \tag{2.8}$$

MIMO antenna systems are expected to offer high-data-throughput wherever they are deployed. To measure the quality of throughput,

Channel Capacity is utilized which defines the maximum transmission of signal over the MIMO channel.

Therefore, without incurring significant error, Channel Capacity Loss (CCL) is calculated which is the upper threshold of data-rate transfer and refers to the continuous transmission of signal irrespective of significant error. Thus CCL is calculated by the following equation

$$CC_{Loss} = -\log_2(\phi^M) \tag{2.9}$$

$$\phi^M = \begin{bmatrix} \varphi_{11} & \varphi_{12} \\ \varphi_{21} & \varphi_{22} \end{bmatrix} \tag{2.10}$$

where

$$\varphi_{11} = 1 - \left[|S_{11}|^2 + |S_{12}|^2\right] \tag{2.11}$$

$$\varphi_{22} = 1 - \left[|S_{22}|^2 + |S_{21}|^2\right] \tag{2.12}$$

$$\varphi_{12} = -\left[S_{11}^* S_{12} + S_{21}^* S_{12}\right] \tag{2.13}$$

$$\varphi_{21} = -\left[S_{22}^* S_{21} + S_{12}^* S_{21}\right] \tag{2.14}$$

For the MIMO configuration system, CCL should be limited to 0.4bits/s/Hz.

Mean Effective gain is the ratio of power absorption of the designed-patch antenna to the mean-power which is incident on MIMO configuration and the reference antenna is considered as an isotropic antenna for calculation of MEG. For consideration of the medium environment and indoor medium, XPR=0db is considered while for outdoor environment XPR=6dB is assumed. Mathematically, MEG is given by

$$MEG = \frac{P_{received}}{P_{incident}}$$

$$= \int_0^{2\pi} \int_0^{2\pi} \left[\frac{X_{PR}}{1 + X_{PR}} F_\theta(\theta,\phi) P_\theta(\theta,\phi) + \frac{1}{1 + X_{PR}} F_\phi(\theta,\phi) P_\phi(\theta,\phi) \right] \sin\theta \, d\theta \, d\phi$$

$$\tag{2.15}$$

Table 2.1 Tabulated diversity performance.

MIMO parameter	Ideal values	Simulated values
ECC	<0.5	<0.005
DG	>9.95dB	>9.975dB
TARC	<0dB	<-30dB
CCL	<0.4b/s/Hz	<0.25b/s/Hz

In the above equation, F_θ, F_ϕ represents power gain patterns of the designed MIMO system. F_θ is calculated by changing θ and keeping ϕ a constant value. Similarly, F_ϕ is calculated by changing ϕ and keeping θ a constant value. Power gain measurement can be carried out by using Dual Ridge Horn Antenna as a radiator. It should be noted that MEG should always lie <-3dB for all bands of operation [12]. Table 2.1 compares the MIMO performance between ideal and simulated values obtained. As per the comparison, the MIMO antenna ensures good diversity performance in all aspects.

2.3   Diversity Performance Analysis of 2×2 UWB-MIMO/Dual-Polarization/UWB: Single, Dual, Triple, and Four Notched Bands

This section discusses the diversity performance of a 2×2 MIMO antenna designed for ultrawideband or multiband applications. Table 2.2 discusses the diversity performance of 2×2 UWB and Multiband MIMO Antenna configuration. The comparison is discussed in terms of size, bandwidth, and diversity characteristics including ECC, DG, and TARC. As per the observations, the multiband antenna [12] provides maximum isolation of more than 25dB for the bandwidth of 4.68-5.75GHz. Also, a compact antenna designed for UWB applications results in isolation of -12dB and -17dB for two different bandwidths.

In general, two perpendicular ports are fed to the radiating patch ensuring polarized modes of excitation. As compared in Table 2.3, dual-polarized antennas are fabricated on Rogers TMM4 and commercially available FR4 substrate. Different isolation techniques are reported to achieve better isolation and a few of them are by etching slots on the patch, using a resonant stub, T-type, and Y-type stub in-ground, and also by using arrow type stub

Table 2.2 Diversity performance comparison of 2×2 UWB and Multiband MIMO Antenna.

Reference	Size of antenna (mm²)	Bandwidth (GHz)	Substrate	Isolation (dB)	Directive gain (dB)	ECC
[1]	$0.53\lambda_o \times 0.56\lambda_o$	4.00-9.00	FR4	<-15	-	-
[2]	$0.25\lambda_o \times 0.241\lambda_o$	2.50-12.0	FR4	<-20	>9.50	<0.02
[5]	$0.212\lambda_o \times 0.299\lambda_o$	2.90-15.0	Kapton	<-15	-	<0.30
[7]	$0.189\lambda_o \times 0.342\lambda_o$	2.70-12.0	Rogers RO4350B	<-15	-	<0.10
[9]	$0.212\lambda_o \times 0.247\lambda_o$	2.20-10.60	FR4	<-12 <-17	- -	- -
[12]	$0.480\lambda_o \times 0.34\lambda_o$	4.68-5.75	F4BK	<-25	-	-
[13]	-	2.447-2.468	F4B	<-20.4	-	-
[14]	$0.458\lambda_o \times 0.458\lambda_o$	4.58-6.12	FR4	<-15	-	<0.15

Table 2.3 Dual polarization MIMO antenna.

Reference	Physical dimension	Isolation technique	Isolation (dB)	Substrate	ECC	Bandwidth (GHz)
[15]	$0.250\lambda_o \times 0.270\lambda_o$	Etching of slots in radiating patch	<-15	Rogers TMM4	<0.01	3.00-11.00
[16]	$0.220\lambda_o \times 0.243\lambda_o$	Resonant stub connected to ground + Etched slots in radiator	<-20	Rogers TMM4	<0.002	3.00-10.60
[17]	$0.400\lambda_o \times 0.400\lambda_o$	T-type stub in ground	<-20	FR4	-	3.00-11.00
[18]	$0.352\lambda_o \times 0.352\lambda_o$	Y-shaped DGS in ground	<-20	FR4	<0.03	3.52-10.66
[19]	$0.350\lambda_o \times 0.350\lambda_o$	Arrow type stub in ground	<-20	FR4	<0.30	3.00-12.00

Table 2.4 UWB single/dual/triple/four notched bands 2×2 MIMO antenna.

Ref.	Size	No. of notched bands/ substrate	Orientation of patch	Isolation technique	Isolation (dB)	ECC	DG (dB)	TARC (dB)
[23]	$0.318\lambda_o \times 0.318\lambda_o$	1 Taconic RF35	Adjacent	Rectangular Stub in ground	<-15	<0.002	-	-
[24]	$0.400\lambda_o \times 0.400\lambda_o$	1 FR4	Orthogonal	Rectangular Stub in ground	<-15	<0.05	-	-
[25]	$0.174\lambda_o \times 0.348\lambda_o$	1	Adjacent	T-type stub in ground	<-20	<0.012	>9.95	-
[26]	$0.\lambda_o \times 0.350\lambda_o$	FR4	Adjacent	T-type stub in ground	<-15	<0.02	-	-
[36]	$0.350\lambda_o \times 0.350\lambda_o$	1	Orthogonal	Rectangular Slotted Stub in ground	<-15	<0.04	>9.60	<-2.50
[39]	$0.350\lambda_o \times 0.350\lambda_o$	FR4	Adjacent	Funnel-type stub in ground	<-24	<0.20	>9.90	<-10

(*Continued*)

Table 2.4 UWB single/dual/triple/four notched bands 2×2 MIMO antenna. (*Continued*)

Ref.	Size	No. of notched bands/ substrate	Orientation of patch	Isolation technique	Isolation (dB)	ECC	DG (dB)	TARC (dB)
[41]	$0.227\lambda_o \times 0.268\lambda_o$	2	Adjacent	T-type slot in the ground	<-20	<0.03	-	-
[43]	$0.206\lambda_o \times 0.372\lambda_o$	2	Adjacent	T-type stub in ground	<-20	<0.03	>9.98	<-30
[45]	$0.2764\lambda_o \times 0.4288\lambda_o$	2	Adjacent	Minkowski fractal in ground + 2 rectangular strips backed plane	<-21.81	<0.01	>9.99	-
[47]	$0.2317\lambda_o \times 0.2949\lambda_o$	3	Adjacent	T-type stub in ground	<-18	<0.02	-	-
[48]	$0.280\lambda_o \times 0.560\lambda_o$	4	Adjacent	Slotted T-type stub attached to the ground	<-20	<0.03	-	-

Table 2.5 Comparison of 4×4 MIMO antenna configuration.

Ref.	Size	Bandwidth (GHz)/ substrate	Isolation technique	Orientation	Isolation (dB)	ECC	DG (dB)	TARC (dB)	CCL (b/s/ Hz)
[51]	$0.848\lambda_o \times 0.848\lambda_o$	3.18-11.50 Taconic RF30	NIL	Orthogonal	<-15	<0.3	>9.98	-	<0.4
[52]	$0.408\lambda_o \times 0.576\lambda_o$	3.52-10.08 FR4	Neutralization	Adjacent	<-23	<0.039	>9.81	<-15	<0.29
[57]	$0.240\lambda_o \times 0.240\lambda_o$	3.00-10.90 FR4	NIL	Orthogonal	<-40	<0.002	-	-	<0.002
[59]	$0.206\lambda_o \times 0.206\lambda_o$	2.00-10.60 FR4	NIL	Orthogonal	<-15	<0.005	>9.97	-	-
[62]	$0.400\lambda_o \times 0.400\lambda_o$	3.00-18.0 FR4	Pair of rectangular slots in ground	Orthogonal	<-20	<0.03	>8.50	-	-

(Continued)

Table 2.5 Comparison of 4×4 MIMO antenna configuration. (*Continued*)

Ref.	Size	Bandwidth (GHz)/ substrate	Isolation technique	Orientation	Isolation (dB)	ECC	DG (dB)	TARC (dB)	CCL (b/s/ Hz)
[66]	$0.420\lambda_o \times 0.420\lambda_o$	3.15-11.36 Rogers RT Duroid 5880	NIL	Orthogonal	<-20	<0.02	>9.998	<-30	-
[68]	$0.516\lambda_o \times 0.516\lambda_o$	3.10-10.40 Rogers RT Duroid 5880	NIL	Orthogonal	<-15	<0.05		-	-
[70]	$0.672\lambda_o \times 0.672\lambda_o$	2.80-13.30 FR4	Interconnected strips in ground	Orthogonal	<-20	<0.06	>9.90	-	-
[72]	$0.530\lambda_o \times 0.530\lambda_o$	3.18-20.10 Rogers RT Duroid 5880	NIL	Orthogonal	<-20	<0.10	>9.99	<-30	-

in the ground. Also, these polarized antennas can be designed not only for UWB applications but also for multiband applications. It can be also noted from the comparison that antenna offers as low as <0.01 ECC with better isolation of more than 20dB.

Table 2.4 shows the diversity comparison of the 2×2 MIMO antenna configuration. As reported, the MIMO antenna provides filtering of single, dual, triple, and four notched bands.

Orientation of the radiating elements plays a major role in achieving isolation between radiating elements along with using any isolating technique. Isolation techniques such as rectangular stub, fractal element etched in-ground, funnel-type stub attached to the ground, T-type stub are used either to improve isolation or to achieve isolation. Table 2.5 signifies the comparison table of the 4×4 MIMO antenna configuration. Also, all the radiating elements are placed either adjacent to each other or they are placed in an orthogonal manner. Reported antennas are fabricated on different substrates available with a compact size as small as 31×31mm².

2.4   4×4 MIMO Antenna

Figure 2.4 shows the arrangement of a 4×4 MIMO antenna configuration placed orthogonal arrangement [51]. All the radiating elements are connected 50Ω SMA connectors via a matched microstrip line.

$$\text{ECC}_{4\times4} = \frac{\left|s_{11}^{\star}s_{12} + s_{21}^{\star}s_{22} + s_{13}^{\star}s_{32} + s_{14}^{\star}s_{42}\right|^2}{\left(\left(1-|s_{11}|^2 - |s_{21}|^2 - |s_{31}|^2 - |s_{41}|^2\right)\left(1-|s_{12}|^2 - |s_{22}|^2 - |s_{32}|^2 - |s_{42}|^2\right)\right)}$$

$$(2.16)$$

Equation 2.16 is used to calculate the ECC of the above MIMO antenna. Also, Figure 2.5 shows the simulation characterization of the 4×4 MIMO antenna. As observed from Figure 2.5(a), all the elements radiate with matched bandwidth of 3.25GHz-11.29GHz. Also, Figure 2.5(b)-(c) shows graphs of transmission coefficients achieved between different ports and are used in Equation 2.16 to calculate the diversity performance of the MIMO antenna. MIMO antenna maintains ECC below 0.005 and DG above 9.99dB in the operating bandwidth of interest.

(a)

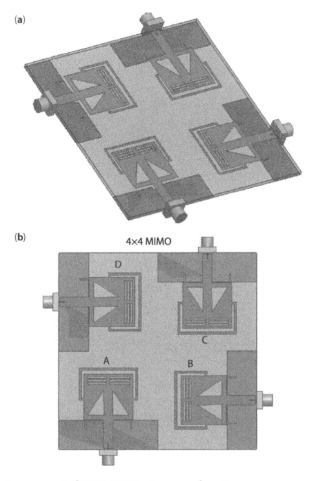

(b)   4×4 MIMO

Figure 2.4 Arrangement of T×T MIMO antenna configuration.

2.5   Conclusions

This chapter focused on the analysis of 2×2 and 4×4 MIMO antenna configurations either working for multiband configuration or UWB/Superwideband applications. Detailed analysis of characterization MIMO diversity performance was studied including envelope Correlation Coefficient, Directive Gain, Total Active Reflection Coefficient, Channel Capacity Loss, and Mean Effective Gain. All the above results were derived by considering 2×2/4×4 MIMO antenna configuration; a reference antenna was used to justify the results and found that these values were well below the allowed permissible values. Also, an extensive comparison is carried out to understand the significance of MIMO performance.

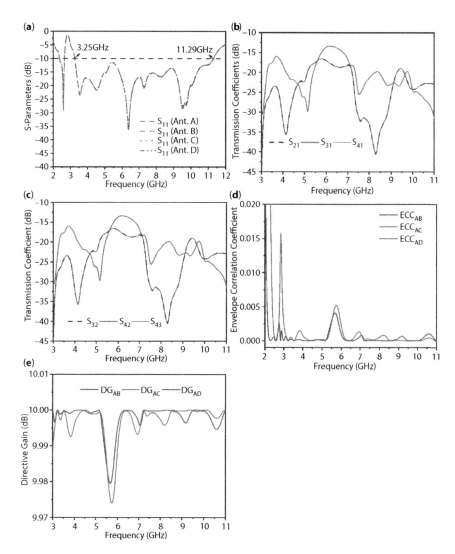

Figure 2.5 Arrangement of 4×4 MIMO antenna configuration (a) S-Paramaters (S_{11}, S_{22}, S_{33}, S_{44}) (b) Transmission Coefficients (S_{21}, S_{31}, S_{41}) (c) Transmission Coefficients (S_{32}, S_{42}, S_{43}) (d) ECC (e) DG.

References

1. T. Meng Guan and S. K. A. Rahim, "Compact monopole MIMO antenna for 5G application," *Microwave and Optical Technology Letters*, vol. 59, no. 5, pp. 1074-1077, 2017.

2. P. C. Nirmal, A. Nandgaonkar, S. Nalbalwar, and R. K. Gupta, "Compact wideband MIMO antenna for 4G WI-MAX, WLAN and UWB applications," *AEU - International Journal of Electronics and Communications*, vol. 99, pp. 284-292, 2019.

3. R. Jian, H. Wei, Y. Yingzeng, and F. Rong, "Compact Printed MIMO Antenna for UWB Applications," *IEEE Antennas and Wireless Propagation Letters*, vol. 13, pp. 1517-1520, 2014.

4. M. A. Ul Haq and S. Koziel, "Ground Plane Alterations for Design of High-Isolation Compact Wideband MIMO Antenna," *IEEE Access*, vol. 6, pp. 48978-48983, 2018.

5. W. Li, Y. Hei, P.M. Grubb, X. Shi and R.T. Chen, "Compact Injet-Printed Flexible MIMO Antenna for UWB Applications," *IEEE Access*, vol. 6, pp. 50290-50298, 2018.

6. M. S. Khan, A.-D. Capobianco, S. M. Asif, D. E. Anagnostou, R. M. Shubair, and B. D. Braaten, "A Compact CSRR-Enabled UWB Diversity Antenna," *IEEE Antennas and Wireless Propagation Letters*, vol. 16, pp. 808-812, 2017.

7. L. Liu, S. W. Cheung, and T. I. Yuk, "Compact multiple-input–multiple-output antenna using quasi-self-complementary antenna structures for ultrawideband applications," *IET Microwaves, Antennas & Propagation*, vol. 8, no. 13, pp. 1021-1029, 2014.

8. Z. Wani and D. K. Vishwakarma, "An ultrawideband antenna for portable MIMO terminals," *Microwave and Optical Technology Letters*, vol. 58, no. 1, pp. 51-57, 2016.

9. Y. Wu and Y. Long, "Compact MIMO antenna for LTE 2500 and UWB applications," *Microwave and Optical Technology Letters*, vol. 57, no. 9, pp. 2046-2049, 2015.

10. J. Tao and Q. Feng, "Compact Ultrawideband MIMO Antenna With Half-Slot Structure," *IEEE Antennas and Wireless Propagation Letters*, vol. 16, pp. 792-795, 2017.

11. M. G. N. Alsath and M. Kanagasabai, "Compact UWB Monopole Antenna for Automotive Communications," *IEEE Transactions on Antennas and Propagation*, vol. 63, no. 9, pp. 4204-4208, 2015.

12. Z. Wang and Y. Dong, "Compact MIMO antenna using Stepped Impedance resonator-based metasurface for 5G and WiFi applications," *Microwave Optical Technology Letters*, pp. 1-6, 2020.

13. G. Zhang and Q. Chen, "Mutual Coupling reduction in Chinese character-shaped artistic MIMO antenna," *Microwave and Optical Technology Letters*, pp. 1-7, 2020.

14. M. Yang and J. Zhou, "A compact diversity MIMO antenna with enhanced bandwidth and high-isolation characteristics for WLAN/5G/WiFi applications," *Microwave and Optical Technology Letters*, pp. 1-12, 2020.

15. M. S. Khan, A.-D. Capobianco, A. Iftikhar, S. Asif, and B. D. Braaten, "A compact dual polarized ultrawideband multiple-input- multiple-output antenna," *Microwave and Optical Technology Letters*, vol. 58, no. 1, pp. 163-166, 2016.

16. M. S. Khan, A.-D. Capobianco, A. Iftikhar, R. M. Shubair, D. E. Anagnostou, and B. D. Braaten, "Ultra-compact dual-polarised UWB MIMO antenna with meandered feeding lines," *IET Microwaves, Antennas & Propagation*, vol. 11, no. 7, pp. 997-1002, 2017.

17. C.-X. Mao and Q.-X. Chu, "Compact Coradiator UWB-MIMO Antenna With Dual Polarization," *IEEE Transactions on Antennas and Propagation*, vol. 62, no. 9, pp. 4474-4480, 2014.

18. J. Zhu, B. Feng, B. Peng, L. Deng, and S. Li, "A dual notched band MIMO slot antenna system with Y-shaped defected ground structure for UWB applications," *Microwave and Optical Technology Letters*, vol. 58, no. 3, pp. 626-630, 2016.

19. J. Zhu, S. Li, B. Feng, L. Deng, and S. Yin, "Compact Dual-Polarized UWB Quasi-Self-Complementary MIMO/Diversity Antenna With Band-Rejection Capability," *IEEE Antennas and Wireless Propagation Letters*, vol. 15, pp. 905-908, 2016.

20. H. Nawaz and I. Tekin, "Dual port disc monopole antenna for wide-band MIMO-based wireless applications," *Microwave and Optical Technology Letters*, vol. 59, no. 11, pp. 2942-2949, 2017.

21. Q. Li, A. P. Feresidis, M. Mavridou, and P. S. Hall, "Miniaturized Double-Layer EBG Structures for Broadband Mutual Coupling Reduction Between UWB Monopoles," *IEEE Transactions on Antennas and Propagation*, vol. 63, no. 3, pp. 1168-1171, 2015.

22. N. K. Sahu, G. Das, and R. K. Gangwar, "Dielectric resonator-based wide band circularly polarized MIMO antenna with pattern diversity for WLAN applications," *Microwave and Optical Technology Letters*, vol. 60, no. 12, pp. 2855-2862, 2018.

23. M. Abedian, S. K. A. Rahim, C. Fumeaux, S. Danesh, Y. C. Lo, and M. H. Jamaluddin, "Compact ultrawideband MIMO dielectric resonator antennas with WLAN band rejection," *IET Microwaves, Antennas & Propagation*, vol. 11, no. 11, pp. 1524-1529, 2017.

24. P. Gao, S. He, X. Wei, Z. Xu, N. Wang, and Y. Zheng, "Compact Printed UWB Diversity Slot Antenna With 5.5-GHz Band-Notched Characteristics," *IEEE Antennas and Wireless Propagation Letters*, vol. 13, pp. 376-379, 2014.

25. R. Chandel and A. K. Gautam, "Compact MIMO/diversity slot antenna for UWB applications with band-notched characteristic," *Electronics Letters*, vol. 52, no. 5, pp. 336-338, 2016.

26. H.-F. Huang and S.-G. Xiao, "Mimo antenna with high frequency selectivity and controllable bandwidth for band-notched UWB applications," *Microwave and Optical Technology Letters*, vol. 58, no. 8, pp. 1886-1891, 2016.

27. A. A. Ibrahim, J. Machac, and R. M. Shubair, "Compact UWB MIMO antenna with pattern diversity and band rejection characteristics," *Microwave and Optical Technology Letters*, vol. 59, no. 6, pp. 1460-1464, 2017.

28. M.S. Khan, A. Iftikhar, R.M. Shubair, A.D. Capobianco, B.D. Braaten and D.E. Anagnostou, "A four element, planar, compact UWB antenna with WLAN

band rejection characteristics," *Microwave Optical Technology Letters*, pp. 1-8, 2020.

29. L. Kang, H. Li, X.-H. Wang, and X.-W. Shi, "Miniaturized band-notched UWB MIMO antenna with high isolation," *Microwave and Optical Technology Letters*, vol. 58, no. 4, pp. 878-881, 2016.

30. M. S. Khan, A. Naqvi, B. Ijaz, M. F. Shafique, B. D. Braaten, and A. D. Capobianco, "Compact planar UWB MIMO antenna with on-demand WLAN rejection," *Electronics Letters*, vol. 51, no. 13, pp. 963-964, 2015.

31. L. Liu, S. W. Cheung, and T. I. Yuk, "Compact MIMO Antenna for Portable UWB Applications With Band-Notched Characteristic," *IEEE Transactions on Antennas and Propagation*, vol. 63, no. 5, pp. 1917-1924, 2015.

32. N. Malekpour, M. Amin Honarvar, A. Dadgarpur, B. S. Virdee, and T. A. Denidni, "Compact UWB mimo antenna with band-notched characteristic," *Microwave and Optical Technology Letters*, vol. 59, no. 5, pp. 1037-1041, 2017.

33. A. Toktas, "G-shaped band-notched ultra-wideband MIMO antenna system for mobile terminals," *IET Microwaves, Antennas & Propagation*, vol. 11, no. 5, pp. 718-725, 2017.

34. S. Tripathi, A. Mohan, and S. Yadav, "A compact octagonal fractal UWBMIMO antenna with WLAN band-rejection," *Microwave and Optical Technology Letters*, vol. 57, no. 8, pp. 1919-1925, 2015.

35. N. Jaglan, B. K. Kanaujia, S. D. Gupta, and S. Srivastava, "Dual Band Notched EBG Structure based UWB MIMO/Diversity Antenna with Reduced Wide Band Electromagnetic Coupling," *Frequenz*, vol. 71, no. 11-12, 2017.

36. Z. Tang, J. Zhan, X. Wu, Z. Xi, L. Chen, and S. Hu, "Design of a compact UWB-MIMO antenna with high isolation and dual band-notched charac-teristics," *Journal of Electromagnetic Waves and Applications*, vol. 34, no. 4, pp. 500-513, 2020.

37. T. Tzu-Chun and L. Ken-Huang, "An Ultrawideband MIMO Antenna With Dual Band-Notched Function," *IEEE Antennas and Wireless Propagation Letters*, vol. 13, pp. 1076-1079, 2014.

38. R. Chandel, A. K. Gautam, and K. Rambabu, "Tapered Fed Compact UWB MIMO-Diversity Antenna With Dual Band-Notched Characteristics," *IEEE Transactions on Antennas and Propagation*, vol. 66, no. 4, pp. 1677-1684, 2018.

39. A. K. Gautam, S. Yadav, and K. Rambabu, "Design of ultra-compact UWB antenna with band-notched characteristics for MIMO applications," *IET Microwaves, Antennas & Propagation*, vol. 12, no. 12, pp. 1895-1900, 2018.

40. H. Huang, Y. Liu, S.-S. Zhang, and S.-X. Gong, "Compact polarization diver-sity ultrawideband mimo antenna with triple band-notched characteristics," *Microwave and Optical Technology Letters*, vol. 57, no. 4, pp. 946-953, 2015.

41. C. R. Jetti and V. R. Nandanavanam, "Trident-shape strip loaded dual band-notched UWB MIMO antenna for portable device applications," *AEU - International Journal of Electronics and Communications*, vol. 83, pp. 11-21, 2018.

42. W. T. Li, Y. Q. Hei, H. Subbaraman, X. W. Shi, and R. T. Chen, "Novel Printed Filtenna With Dual Notches and Good Out-of-Band Characteristics for UWB-MIMO Applications," *IEEE Microwave and Wireless Components Letters*, vol. 26, no. 10, pp. 765-767, 2016.

43. M. Sharma, "Design and Analysis of MIMO Antenna with High Isolation and Dual Notched Band Characteristics for Wireless Applications," *Wireless Personal Communications*, vol. 112, no. 3, pp. 1587-1599, 2020.

44. N. Jaglan, S. D. Gupta, E. Thakur, D. Kumar, B. K. Kanaujia, and S. Srivastava, "Triple band notched mushroom and uniplanar EBG structures based UWB MIMO/Diversity antenna with enhanced wide band isolation," *AEU - International Journal of Electronics and Communications*, vol. 90, pp. 36-44, 2018.

45. J. Banerjee, A. Karmakar, R. Ghatak, and D. R. Poddar, "Compact CPW-fed UWB MIMO antenna with a novel modified Minkowski fractal defected ground structure (DGS) for high isolation and triple band-notch characteristic," *Journal of Electromagnetic Waves and Applications*, vol. 31, no. 15, pp. 1550-1565, 2017.

46. Y. Kong, Y. Li, and K. Yu, "A Minimized MIMO-UWB Antenna with High Isolation and Triple Band-Notched Functions," *Frequenz*, vol. 70, no. 11-12, 2016.

47. Z.-X. Yang, H.-C. Yang, J.-S. Hong, and Y. Li, "A miniaturized triple band-notched MIMO antenna for UWB application," *Microwave and Optical Technology Letters*, vol. 58, no. 3, pp. 642-647, 2016.

48. L. Wu, H. Lyu, H. Yu, and J. Xu, "Design of a Miniaturized UWB-MIMO Antenna with Four Notched-Band Characteristics," *Frequenz*, vol. 73, no. 7-8, pp. 245-252, 2019.

49. B. T. Ahmed, P. S. Olivares, J. L. M. Campos, and F. M. Vázquez, "(3.1– 20) GHz MIMO antennas," *AEU - International Journal of Electronics and Communications*, vol. 94, pp. 348-358, 2018.

50. M. Shehata, M. S. Said, and H. Mostafa, "Dual Notched Band Quad-Element MIMO Antenna With Multitone Interference Suppression for IR-UWB Wireless Applications," *IEEE Transactions on Antennas and Propagation*, vol. 66, no. 11, pp. 5737-5746, 2018.

51. M. N. Hasan, S. Chu, and S. Bashir, "A DGS monopole antenna loaded with U-shape stub for UWB MIMO applications," *Microwave and Optical Technology Letters*, vol. 61, no. 9, pp. 2141-2149, 2019.

52. R. N. Tiwari, P. Singh, B. K. Kanaujia, and K. Srivastava, "Neutralization technique based two and four port high isolation MIMO antennas for UWB communication," *AEU - International Journal of Electronics and Communications*, vol. 110, 2019.

53. W. A. E. Ali and A. A. Ibrahim, "A compact double-sided MIMO antenna with an improved isolation for UWB applications," *AEU - International Journal of Electronics and Communications*, vol. 82, pp. 7-13, 2017.

54. T. Dabas, D. Gangwar, B. K. Kanaujia, and A. K. Gautam, "Mutual coupling reduction between elements of UWB MIMO antenna using small size uniplanar EBG exhibiting multiple stop bands," *AEU - International Journal of Electronics and Communications*, vol. 93, pp. 32-38, 2018.

55. R. Mathur and S. Dwari, "Compact CPW-Fed ultrawideband MIMO antenna using hexagonal ring monopole antenna elements," *AEU - International Journal of Electronics and Communications*, vol. 93, pp. 1-6, 2018.

56. D. Sipal, M. P. Abegaonkar, and S. K. Koul, "Compact planar 2×2 and 4×4 UWB mimo antenna arrays for portable wireless devices," *Microwave and Optical Technology Letters*, vol. 60, no. 1, pp. 86-92, 2018.

57. G. Srivastava, A. Mohan, and A. Chakraborty, "A compact multidirectional UWB MIMO slot antenna with high isolation," *Microwave and Optical Technology Letters*, vol. 59, no. 2, pp. 243-248, 2017.

58. Z. Wani and D. Kumar, "A compact 4×4 MIMO antenna for UWB applications," *Microwave and Optical Technology Letters*, vol. 58, no. 6, pp. 1433-1436, 2016.

59. B. Yang, M. Chen, and L. Li, "Design of a four-element WLAN/LTE/UWB MIMO antenna using half-slot structure," *AEU - International Journal of Electronics and Communications*, vol. 93, pp. 354-359, 2018.

60. S. Kumar, R. Kumar, R. Kumar Vishwakarma, and K. Srivastava, "An improved compact MIMO antenna for wireless applications with band-notched characteristics," *AEU - International Journal of Electronics and Communications*, vol. 90, pp. 20-29, 2018.

61. S. M. Khan, A. Iftikhar, S. M. Asif, A.-D. Capobianco, and B. D. Braaten, "A compact four elements UWB MIMO antenna with on-demand WLAN rejection," *Microwave and Optical Technology Letters*, vol. 58, no. 2, pp. 270-276, 2016.

62. Z. Tang, J. Zhan, X. Wu, Z. Xi, and S. Wu, "Simple ultra-wider-bandwidth MIMO antenna integrated by double decoupling branches and square-ring ground structure," *Microwave and Optical Technology Letters*, vol. 62, no. 3, pp. 1259-1266, 2019.

63. M. M. Hassan *et al.*, "A novel UWB MIMO antenna array with band notch characteristics using parasitic decoupler," *Journal of Electromagnetic Waves and Applications*, vol. 34, no. 9, pp. 1225-1238, 2019.

64. S. Tripathi, A. Mohan, and S. Yadav, "A Compact Koch Fractal UWB MIMO Antenna With WLAN Band-Rejection," *IEEE Antennas and Wireless Propagation Letters*, vol. 14, pp. 1565-1568, 2015.

65. R. Gomez-Villanueva and H. Jardon-Aguilar, "Compact UWB Uniplanar Four-Port MIMO Antenna Array With Rejecting Band," *IEEE Antennas and Wireless Propagation Letters*, vol. 18, no. 12, pp. 2543-2547, 2019.

66. V. Dhasarathan, T. K. Nguyen, M. Sharma, S. K. Patel, S. K. Mittal, and M. T. Pandian, "Design, analysis and characterization of four port multiple-input-multiple-output UWB-X band antenna with band rejection ability

for wireless network applications," *Wireless Networks*, vol. 26, no. 6, pp. 4287-4302, 2020.

67. G. Kan, W. Lin, C. Liu, and D. Zou, "An array antenna based on coplanar parasitic patch structure," *Microwave and Optical Technology Letters,* vol. 60, no. 4, pp. 1016-1023, 2018.

68. H. Liu, G. Kang, and S. Jiang, "Compact dual band-notched UWB multiple-input multiple-output antenna for portable applications," *Microwave and Optical Technology Letters*, vol. 62, no. 3, pp. 1215-1221, 2020.

69. P. Bactavatchalame and K. Rajakani, "Compact broadband slot-based MIMO antenna array for vehicular enviornment," *Microwave Optical Technology Letters*, pp. 1-9, 2020.

70. S. Kumar, G. H. Lee, D. H. Kim, W. Mohyuddin, H. C. Choi, and K. W. Kim, "Multiple-input-multiple-output/diversity antenna with dual band-notched characteristics for ultra-wideband applications," *Microwave and Optical Technology Letters*, vol. 62, no. 1, pp. 336-345, 2019.

71. A. Gorai, A. Dasgupta, and R. Ghatak, "A compact quasi-self-complementary dual band notched UWB MIMO antenna with enhanced isolation using Hilbert fractal slot," *AEU - International Journal of Electronics and Communications*, vol. 94, pp. 36-41, 2018.

72. M. Sharma, V. Dhasarathan, S. K. Patel, and T. K. Nguyen, "An ultra-compact four-port 4 × 4 superwideband MIMO antenna including mitigation of dual notched bands characteristics designed for wireless network applications," *AEU - International Journal of Electronics and Communications*, vol. 123, 2020.

73. D. K. Raheja, B. K. Kanaujia, and S. Kumar, "Low profile four-port superwideband multiple-input-multiple-output antenna with triple band rejection characteristics," *International Journal of RF and Microwave Computer-Aided Engineering*, vol. 29, no. 10, 2019.

74. D.J. Lee, S.J. Lee, S.T. Khang, and J.W. Wu, "Extensible compact 8-Port MIMO antenna with pattern gain," *Microwave Optical Technology Letters*, vol. 59, no. 2, pp. 236-240, 2017.

75. J.Y. Deng, J. Yao, D.Q. Sun and L.X. Guo, "Ten-element MIMO antenna for 5G terminals," *Microwave Optical Technology Letters*, vol. 60, pp. 3.45-3.49, 2018.

3

Various Antenna Array Designs Using Scilab Software: An Exploratory Study

V. A. Sankar Ponnapalli* and Praveena A

Department of Electronics and Communication Engineering, Sreyas Institute of Engineering and Technology, Hyderabad, India

Abstract

This study illustrates the use of open-source software called Scilab for the design of conventional and advanced antenna arrays. Array factor codes and simulation results of various array antennas such as circular, concentric circular, square, and hexagonal antenna arrays are demonstrated in this work. This study revealed that Scilab is a good alternative to commercial software, and an effective electromagnetic, and antenna teaching aid for online, blended and distance mode. Teaching the concepts like antenna arrays is a challenging task for the teacher owing to the various equations involved to demonstrate antenna array performance, and the simulation tool is needed to easily show the array factor properties to the students. But the use of commercial software is expensive and students cannot work outside of the college. Therefore Scilab free and open-source software for array antennas have been demonstrated here and it is a good alternative to commercial software.

Keywords: Scilab, antenna arrays, teaching, open source, array factor, engineering

3.1 Introduction

Antenna theory is one of the classic subjects in electrical communication engineering. This course is usually offered in the third year of the electronics and communication engineering program, and the prerequisite of this course is electromagnetic theory. The learning outcomes for the antenna

---

Corresponding author: vadityasankar3@gmail.com

Prashant Ranjan, Dharmendra Kumar Jhariya, Manoj Gupta, Krishna Kumar, and Pradeep Kumar (eds.) Next-Generation Antennas: Advances and Challenges, (49–60) © 2021 Scrivener Publishing LLC

engineering course are given in Table 3.1. An antenna engineering course requires a strong mathematical foundation of vector calculus, differential, and integral equations. However, this course provides strong insights into wireless communication and advanced wireless systems such as mobile, satellite, and radar communications [1–4]. The radiation pattern behavior of antenna systems depends only on the geometrical characteristics of the antenna. Simulation tools are therefore the best teaching aids for the antenna engineering courses, and students can easily understand radiation properties through simulation outputs.

The antenna array topic is a part of the antenna theory course, which emphasizes the impact of antenna array geometry and feeding on the radiation pattern. In general, at undergraduate-level courses, various linear array antennas ranging from the two elements to the n-element arrays are discussed. Linear, circular and planar array antennas are part of the curriculum in postgraduate courses. Antenna arrays are mostly suitable for radar, and long-distance communications owing to their beamsteering capability. Simulation tools can easily demonstrate beamforming and beamsteering properties and help students to understand these concepts perfectly [5–7]. The usage of simulation tools also develops the computational thinking skills of the students, which motivates the students in the direction of antenna array research. The design, analysis, and optimization of antenna arrays have been a common research activity over the past few

Table 3.1 Learning Outcomes of antenna engineering course.

S. no.	Learning Outcomes (LO's)
LO:1	Students will be able to learn the very basic, ideal, and conventional antennas like isotropic, monopole, halfwave dipole, and loop antennas.
LO:2	Students will be able to analyse radiation mechanisms and apply Maxwell's equations to antennas.
LO:3	Students will be able to understand the concepts of antenna arrays, types of antenna arrays, beamsteering, and beamforming mechanism of antenna arrays.
LO:4	Students will be able to analyse various microwave antennas and their radiation properties: Yagi-Uda, Horn, Hellical, Dish, Microstrip patch, and whip antennas.
LO:5	Students will be able to calculate and measure the antenna properties, and also understand wave propagation concepts.

decades, and this is evident from the various well-known digital libraries such as the IEEE digital library, ScienceDirect, Springer, and ACM digital library. Due to the advancements of communication systems, there is still research potential in this area. Usage of simulation tools in the classroom teaching influences students and motivates them in the direction of core research-related aspects. But usage of commercial software is again a burden on academic institutions, and students cannot work within college time limits. The best way to bridge these gaps is to use open-source software, and in this book chapter Scilab open-source software is used to develop the codes for array antennas [8–10], and the detailed description of Scilab is presented in this chapter. Various antenna array descriptions, respective Scilab codes, and the simulation results have also been presented. The final discussion of the book chapter is presented as summary remarks in the concluding section.

3.2   Scilab: An Open-Source Software Solution

This software is a liberated, unlocked, cross-stage computational package, and a high stage programming. It provides a translated system environment, with metrics as a form of big data. By this matrix-dependent calculation, vibrant typing, and involuntary memory organization, several numerical problems can be recognized by a decreased number of code lines, contrasted to parallel solutions using native languages, such as Fortran, C, or C ++. This permits clients to rapidly construct models for an array of mathematical problems. This open-source software package also offers a collection of high-level missions such as integration and intricate multidimensional computations. This open-source software can be used for various fields of science and technology such as signal processing, statistics, image processing techniques, fluid dynamic relationships, and optimization techniques. This software also features a free package called Xcos, and it is an open-source counterpart to a commercial software package Simulink from MathWorks. This software syntax is similar to the licensed software Matlab, and it also includes a source code translator to convert code from Matlab to Scilab. This software package is available freely under an open-source accreditation. Owing to the open nature of the software, some user contributions are included in the advanced program. Scilab has many toolboxes that offer a variety of functions, such as, image processing toolbox, wavelet toolbox, optimization toolbox, Scilab Remote Access Module, and some more advanced toolboxes for the engineering and science applications [11–13]. The evolution of Scilab is represented in Figure 3.1.

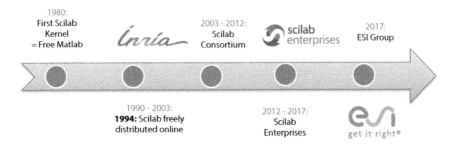

Figure 3.1 Evolution of Scilab [11].

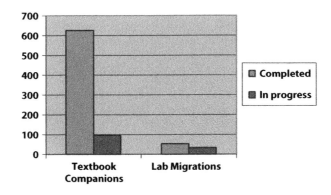

Figure 3.2 Statistics of Scilab usage in India [14].

Scilab is available under a general public license (GPL), where users can use the software freely for any purpose, customize the software to user needs, share the software with peers, and change the code. Owing to the GPL, this software will be useful for the computation thinking among the k12 students, and all streams of engineering, and science educational programs. The usage of Scilab in academic institutions of India is also increasing in the form of lab migrations, and textbook companions as depicted in Figure 3.2 [14–16]. Scilab applied knowledge majorly covers the scientific fields such as mathematics, optimization, statistics, signal and image processing, and control systems.

3.3   Antenna Array Design Using Scilab: Codes and Results

Scilab codes of the circular, concentric circular, square, and hexagonal antenna arrays are discussed in this chapter and the corresponding codes have been presented in Figures 3.3, 3.4, 3.5, and 3.6, respectively. To develop

(a)

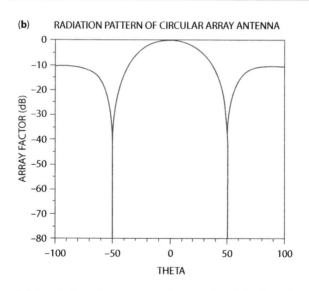

```
File Edit Format Options Window Execute ?
EXP01.sce (D:\SPEYA\Svesearch papers\wiley-sp book chapter\final book\EXP01.sce) - SciNotes
EXP01.sce
 1  //1.RADIATION PATTERN (RECTANGULAR PLOT) OF CIRCULAR ARRAY ANTENNA
 2  clc;
 3  clear;
 4  close;
 5  r=input('radius=');// IN WAVELENGHTS
 6  N=input('NUMBER OF ANTENNA ELEMENTS=')
 7  af=0;
 8  thio=0; //STEERING ANGLE
 9  phio=0; // STEERING ANGLE
10  phi=%pi/2;
11  thi=-%pi:%pi/10000:%pi;
12  k=2*%pi
13      for n=1:N // NUMBER OF ANTENNA ELEMENTS
14          phin=(n-1).*(2.*%pi/N)
15          a=(exp(%i.*k.*r.*(((sin(thi)*cos(phi-phin)))-(sin(thio)*cos(phio-phin))))));
16          af=a+af;
17      end
18  af1=abs(af);
19  af1=af1/max(max((af1)));
20  CAA=20.*log10(af1);
21  plot((thi*57.3),(CAA));
22  xlabel('THETA');
23  ylabel('ARRAY FACTOR (dB)');
24  title('RADIATION PATTERN OF CIRCULAR ARRAY ANTENNA');
25  h=gca();
26  h.data_bounds=[-90,-80;90,0];
27
28  // INPUT PARAMETERS
29  //radius=0.5
30  //NUMBER OF ANTENNA ELEMENTS=10
```

(b)   RADIATION PATTERN OF CIRCULAR ARRAY ANTENNA

Figure 3.3 (a) Scilab code for radiation pattern (rectangular plot) of circular array antenna (b) Corresponding array factor.

(a)

(b)

RADIATION PATTERN OF CIRCULAR ARRAY ANTENNA

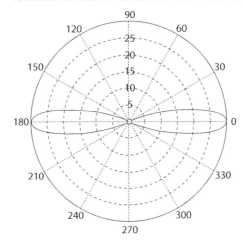

Figure 3.4 (a) Scilab code for radiation pattern (polar plot) of concentric circular array antenna (b) corresponding array factor.

the codes for any type of antenna arrays, the array factor formula of the considered array [3] is required and the various parameters included in the array factor equation must be defined. In this work, circular antenna arrays with 10 elements, and a radius of 0.5λ were considered. Corresponding

(a)

(b)

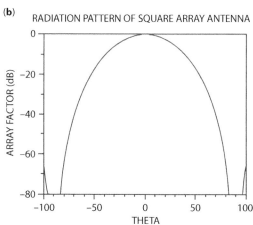

Figure 3.5 (a) Scilab code for radiation pattern (rectangular plot) of square array antenna (b) corresponding array factor.

array factor plots in the rectangular coordinates are represented in Figure 3.3(b), sidelobe level of -10dB, the half power beamwidth of 41.8°, and side lobe level angle of 92.2° has been achieved for this array. Concentric circular antenna arrays consist of 10 elements in each concentric circle with radii of 0.5 λ, λ, and 1.5 λ respectively. Sidelobe level of -28.8dB, half-power beamwidth of 19.2°, and side lobe level angle of 31.9° has been achieved

(a)

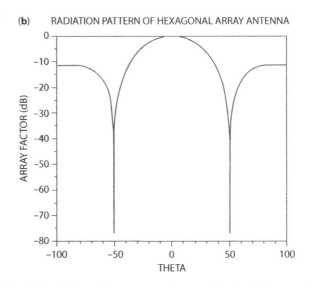

```
//4.RADIATION PATTERN (RECTANGULAR PLOT) OF HEXAGONAL ARRAY ANTENNA

clc;
clear;
clcwr;
af=0;
thio=0; //STEERING ANGLE
phio=0; // STEERING ANGLE
phi=%pi/2;
thi=-%pi:%pi/10000:%pi;
r=0.5;//radius
k=2*%pi
    for n=1:6 // NUMBER OF ANTENNA ELEMENTS
        phin=(n-1).*(2.*%pi/6);
        a=(exp(%i.*k.*r.*(((sin(thi)*cos(phi-phin)))-(sin(thio)*cos(phio-phin)))));
        af=a+af;
    end
afl=abs(af);
afl=afl/max(max((afl)));
HAA=20.*log10(afl);
plot((thi*57.3),(HAA));
xlabel('THETA');
ylabel('ARRAY FACTOR (dB)')
title('RADIATION PATTERN OF HEXAGONAL ARRAY ANTENNA');
h=gca();
h.data_bounds=[-90,-80;90,0];
```

(b)   RADIATION PATTERN OF HEXAGONAL ARRAY ANTENNA

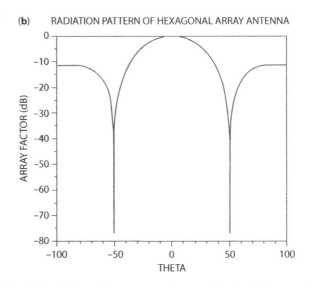

Figure 3.6 (a) Scilab code for radiation pattern (rectangular plot) of hexagonal array antenna (b) corresponding array factor.

for concentric circular arrays as depicted in Figure 3.4(b). A four-element square array antenna is also considered with a sidelobe level of -0.47, the half-power beamwidth 43.4°, and the sidelobe level angle of 171.1° have been achieved as shown in Figure 3.5(b). Finally, the hexagonal antenna

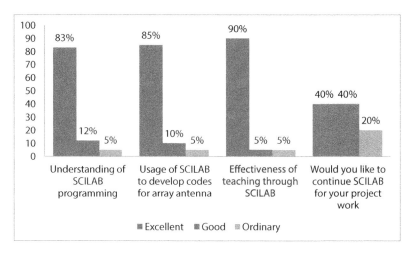

Figure 3.7 Feedback from the students on Scilab programming.

array has been investigated in this chapter, and the array factor of this array is depicted in Figure 3.6(b).

In the context of outcome-based education, the program outcomes (POs) to list PO:1, PO:2, PO:3, PO:4, and PO5 has been strongly attained due to the usage of Scilab coding in the class or laboratory teaching. This open-source software is the best solution for online or remote teaching methods. Feedback from the students of the third-year electronics and communication engineering branch of sreyas IET on the Scilab-aided teaching for the antenna course during the COVID-19 epidemic is depicted in Figure 3.7. It revealed that students are interested in this open-source software-based teaching and some students want to pursue their major level projects with Scilab programming.

3.4   Conclusions

Different antenna array designs using Scilab programming are discussed in this chapter. The teaching antenna array is a challenge for faculty members owing to the various design equations involved in it. Getting maximum attention from students is another challenge in online class/remote teaching mode. However, beam steering concepts of the antenna array can easily be understood by the students through simulated results. Scilab open-source software-based teaching is a key solution to bridge these gaps, and students will also get hands-on experience due to the open and free availability of the software. In this book chapter nearly four different types of antenna arrays, namely circular, concentric circular, square, and hexagonal antenna arrays have been considered and analyzed using the Scilab. All array antenna codes presented in the chapter are implemented, and the simulated outputs are the same as commercial software outputs. The coding instructions for implementing the antenna arrays discussed are also very simple. There has been a positive response from the students to the use of Scilab software in classroom teaching, and Scilab software has strongly attained program outcome which measures the "advanced tool usage". The computational thinking capability of the students will also increase towards core problems owing to these types of teaching practices.

References

1. Ülker Sadık. Antennas and propagation course in education. *International Journal of Electrical Engineering & Education*, Vol. 2018, pp. 1-20, 2018.
2. John. D. Kraus and Ronald .J. Marhefka. *Antennas: For All Application*. 3rd ed. New York: McGraw-Hill, 1997.
3. Balanis CA. *Antenna Theory: Analysis and Design*. 3rd ed. New York: John Wiley, 2005.
4. V. A. Sankar Ponnapalli and P. V. Y. Jayasree. Design of multi-beam rhombus fractal array antenna using new geometric design methodology. *Progress in Electromagnetics Research C*, Vol. 64, pp. 151-158, 2016.
5. M. Rodríguez, V. González, J. E. González, C. Rueda, L. De Haro & C. Martín-Pascual. Development of educational software for the teaching of telecommunication engineering by using MATLAB. *European Journal of Engineering Education*, Vol. 26, pp. 361-374, 2001.
6. R. J. Mailloux. *Phased Array Antenna Handbook*. Artech House, Norwood, MA, 2005.
7. Giovanni Tartariniv *et al*. Consolidating the Electromagnetic Education of Graduate Students Through an Integrated Course. *IEEE Transactions on Education*, Vol. 56, pp. 416-423, 2013.
8. Fedeli, Alessandro, Claudio Montecucco, and Gian Luigi Gragnani. Free and Open Source Software Codes for Antenna Design: Preliminary Numerical Experiments. *Scientific Journal of Riga Technical University-Electrical, Control and Communication Engineering*, Vol. 15, pp. 88-95, 2019.
9. Ponnapalli, VA Sankar, and Vinod Babu. "A Study on Scilab Free and Open Source Programming for Antenna Array Design." *2020 IEEE International IOT, Electronics and Mechatronics Conference (IEMTRONICS)*. IEEE, 2020.
10. Ali El-Hajj, Karim Y. Kabalan, and Mohammed Al-Husseini. Antenna Array Design Using Spreadsheets. *IEEE Transactions on Education*, Vol. 46, pp. 319-324, 2003.
11. https://www.scilab.org/. Accessed on 2020/02/03.

12. Baudin, Michael, and Serge Steer. "Optimization with scilab, present and future." 2009 IEEE International Workshop on Open-source Software for Scientific Computation (OSSC). IEEE, 2009.
13. Vieira EB, Busch WF, Prata DM, Santos LS. Application of Scilab/Xcos for process control applied to chemical engineering educational projects. *Comput Appl Eng Educ.*, Vol. 27, pp. 154–165, 2019.
14. https://scilab.in/. Accessed on 2020/02/03.
15. Jagdish Y. Patil, Balashish Dubey, Kannan, M. Moudgalya, Rakesh Peter. GNURadio, Scilab, Xcos and COMEDI for Data Acquisition and Control: An Open Source Alternative to LabVIEW. In: *8th IFAC Symposium on Advanced Control of Chemical Processes The International Federation of Automatic Control, Singapore, July 10-13, 2012.*
16. Kotha Manohar, Koppuravuri Sravani, V. A. Sankar Ponnapalli. "An investigation on scilab software for the design of transform techniques and digital filters." *2021 International Conference on Computer Communication and Informatics (ICCCI -2021), Jan. 27–29, 2021, Coimbatore, India.* IEEE, 2021.

4

Conformal Wearable Antenna Design, Implementation and Challenges

Brajlata Chauhan[1], Vivek Kumar Srivastava[1], Amrindra Pal[1]* and Sandip Vijay[2]

[1]Microwave and Antenna Lab, Department of EECE, DIT University, Dehradun, Uttarakhand, India
[2]Department of EECE, Shivalik College of Engineering, Dehradun, Uttarakhand, India

Abstract

The demand for wearable devices and systems has risen over the last decade. Wearable antennas are employed for various applications, mostly on and inside wireless body field networks (WBAN) such as wristwatches, jackets, helmets, electronics toys, biomedical systems, etc. The proposed work aims to critically examine the conformal singly curve structure, planar, and conformal slotted structure at 4.6 GHz. A wearable conformal antenna has been developed for resonating frequency between 7.5GHz to 10.5Ghz for different textile substrates. Rogers 3850- an LCP substrate material that has a dielectric constant εr of 2.9, loss tangent $\tan\delta$ of 0.0025 and thickness t = 1.1 mm is used for designing. Further, scattering parameter S11 and radiation characteristics have been analyzed for different substrates as Polyester, Polyethylene, Polyfloncuflon (tm), Polystyrene, and Polyamide. These substrates are applied as new materials in medical systems used to measure human body parameters, and innovative designs such as miniaturized button, multi-band antennas for smart implementations in WBAN. Wearable conformal antenna with parameter optimization simulated on ANSYS High-Frequency Structural Simulator (HFSS) program designs the antenna and structure is conformed on CST microwave studio. The proposed Antenna scale is 9mm x 5.69mm x 1.1mm and the incorporated guard line has a footprint of 5.65mm x 1.2mm. The properties of antenna-like return loss, VSWR and radiation characteristics are studied for the antenna designed.

Keywords: Conformal antenna, wearable antenna, EM radiation, wireless body area network, defected ground structure

*Corresponding author: amrindra.ieee@gmail.com

Prashant Ranjan, Dharmendra Kumar Jhariya, Manoj Gupta, Krishna Kumar, and Pradeep Kumar (eds.) Next-Generation Antennas: Advances and Challenges, (61–90) © 2021 Scrivener Publishing LLC

4.1   Introduction

Antennas are one of the most essential components of wireless communication systems; they radiate electromagnetic energy in the desired directions and are used to receive or radiate the electromagnetic energy. Antennas can be employed not only in space or air but can also be operated underwater, inside and on the human body, on the surface of any shape of radome or even through soil and rock at kHz range of frequencies for short distances [1–3]. The fields of modern antennas have drawn the attention of scientists, researchers and academicians in the last two decades. A high-performance wearable antenna may develop with some trade-off in terms of characteristics parameters with its radiation parameter, specific absorption rate (SAR) and low backward radiation with a less powerful effect on the human body. A textile material may be considered for optimizing the antenna gain, bandwidth and low SAR suggested by the Federal Communications Commission (FCC) [4–6].

Wearable conformal antenna with superior performance and versatile features can be useful where approaches of remarkable antenna performance are necessary for numerous applications. The theory includes the antenna-integration on curved surfaces for more extensive inclusion, for military aircraft or media/military vehicles, smartwatches, etc., in which the antenna is integrated into the structure itself and the systems are integrated for the wearable system network which is applicable for textile antennas, body-worn and biomedical-antenna [7–12].

An important characteristic of a conformal antenna array is the radiation pattern in response to the excitations of its antenna elements. While confirming the antenna structure its radiation properties start to degrade. To improve the overall array performance of an antenna, it is conceivable to change the characteristics and revert the radiation patterns to their original position. The conformal antenna can be introduced for a few aerodynamic systems to reduce aerodynamic drag. Conformal antennas include the phased-array antenna class that comprises an array of planar microstrip patch antennas and dipoles antenna surface covering [1–3, 13–15].

Alongside these antennas exhibit the radiation characteristic Omnidirectional, broadband, miniaturization, multiband characteristics like a phased array.

A Wearable Antenna system is required to work for:

   i)   Greater radiation efficiency at the desired frequency
   ii)   Wideband/ultra-wideband operation
   iii) High gain & bandwidth product and beam steering

iv)   Beamwidth and side lobe level
v)   Specific Absorption Rate calculations for human tissue
vi)   Capabilities of progressively adaptive beam-forming
vii)   Cost-effectiveness and Structural robustness

The radiation pattern/performance may be enhanced by using the change in shape; work includes appropriate active and passive mechanical steering/dampening phase compensation network and voltage amplitude [3, 14].

4.2   Conformal Antenna

Conformal antennas belong to the class of phased array antennas. The most important application of conformal antenna is found in aviation aircraft. It has supplementary extensive applications in defense as well as civilian systems. It has a varied range of types and geometrical structures. The cylindrical surface is the most preferred geometry in the field of microstrip conformal antenna array because of its huge application and a lot of the references are available in the related geometry [1–3, 13–14]. The second preferred geometry is the sphere for the conformal antenna. In a conformal antenna, a specific angle is bent within a restricted space. The angle is bent in such a precise manner that the performance of the antenna remains unaffected. Some of the basics of conformal antenna have been briefly described here to highlight the concepts of the conformal array antenna.

4.2.1   Singly Curved Surfaces

Singly curved conformal antenna constitutes a significant class where high azimuthal angular exposure is required. Such styles of antenna have been applied in countless applications such as medical and communication systems. Basic aperture antennas such as flush-mounted circular cylindrical conductors present an important class of conformal antennas. A single layer curved surface can also be applied as an estimate of the size and shape of body part knee, soldier and indigestive capsule. An orientation insensitive antenna is investigated for the wireless capsule endoscopy system at 2.4 GHz. Polarization diversity is achieved through the three orthogonal currents by bending a dipole [14]. The recently developed antenna is a ring-shape capsule endoscope antenna that operates at the 2.4 GHz ISM band. Other ISM band frequencies are 5150–5350 MHz and 5725–5875 MHz, cordless telephone is available and has a large installation base [16].

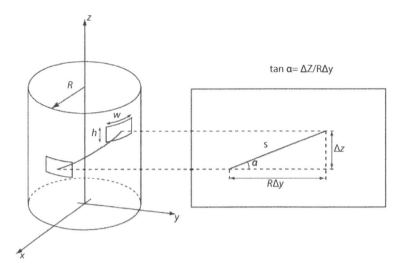

Figure 4.1 Structure of planar microstrip patch antenna.

The meandering dipole structure is applied as the radiator to optimize the antenna size.

In Figure 4.1, an aperture that has a rectangular shape may place on either circumferential or axial orientation. Cylindrical radius is represented by R. The aperture is demonstrated by the patch width w and height h and S is the geodesic length between the centers of the apertures. It is excited by the fundamental TE10 mode gives linearly polarized orientation. For the axial orientation apertures, in which w is less than h, represents the case with circumferential polarization (horizontal). Cross-polarized cases may also consider, that is coupling with one aperture of each type [14].

4.3   Characteristics of Conformal Antenna

4.3.1   Radiation Pattern

The radiation pattern is the characteristic of an antenna and can be deduced by the computation methodology. An antenna may produce radiation characteristics as directive or omnidirectional patterns, shaped beams, multi-beam, main lobe, low side-lobes and so on—the mentioned performance parameter can be obtained with different arrays types. However, the radiation patterns have degraded from planar to conformal transformation and some other design-related issues to consider in the conformal antenna.

Some special techniques like Method of Moment (MoM) and finite difference time domain (FDTD) are also used for analyzing conformal arrays and predicting their performance.

4.3.2   Scan-Invariant Pattern

Conformal antenna arrays, which have rotational symmetry show scan-invariant patterns in one direction (at least). In the array, each element may be excited with different excitation to move the radiation pattern in the desired direction. The Scanning by commutating is applicable on curvature surface. This process needs radiating elements' phase control, to compensate for the effect of array curvature—the required compensation changes with the beam direction.

4.3.3   Phase-Scanned Pattern

If the active region is set, the beam can be shifted by controlling the phase of the component excitations, as in the planar arrays. But the radiating elements on the other side of the active region that is away from the preferred radiation direction with the rising scan angle expect the pattern form to differ significantly with the direction of the beam [44]. When there is a proper phase compensation, side-lobes take place. As the orientation of the scan increases, the side lobe rises and the gain decreases. As the scan-angle increases, the side lobe rises and the gain decreases. To achieve better performance, it seems reasonable to turn off the elements that contribute very poorly, or at least to reduce their amplitudes.

4.3.4   Polarization

The radiation function can be branched into a (vector) element factor and a (scalar) array factor for ordinary planar arrays. The polarisation facts are included in the radiating elements, as polarisation control at the elementary level has become essential [4]. The curvature of the antenna structure will give rise to cross polarisation, even though the radiating elements themselves are free from cross polarisation. The cross polarisation contributions from various elements will also differ if the curvature varies through the array. However, it is possible to synthesize the radiation pattern for low cross polarisation with individual control of two orthogonal radiating components at the element level [17]. For an arc or ring array, the radiation function can be written in terms of 8 and 12 components where the amplitudes depend on scalar expressions drawn in and derivatives of a normalized array factor. Therefore, the designs of cylindrical, conical, and spherical arrays can be constructed from an arc pattern superposition [4].

4.4   Design Methodology - Antenna Modeling

This section is about the design methodology of the proposed wearable antenna which contains the design calculation of the patch and slot of defective ground. Further, the simulation analysis of the line parameter and the radiation characteristics has been carried out on HFSS and CST MWS software.

4.4.1   Overview

The approach used for the proposed antenna contains the numerical analysis where the elementary parameters have been calculated using microstrip transmission line theory. The simulated results of the designed structure are distinguished from the calculated and theoretical results. The antenna structure has been optimized as per the desired resonant frequency and other radiation parameters so that it can meet up to theoretical results. Further, the analysis of the overall simulation results of the proposed wearable antenna has been discussed and presented in Section 4.5.

4.4.2   Geometry and Calculation of Planar MSA

Design of planar microstrip patch antenna having U- and E-shape slot cut in the patch and then conformed on a cylindrical surface has been presented here.

The microstrip patch is lightweight, simple to fabricate and conform and the limitations of it are narrow bandwidth and low gain. The geometry of the planar microstrip patch antenna having E- and U-shape slot is given in Figure 4.2.

The operating frequency of the designed slotted antenna is 4.6 GHz. By applying some basic formulae of patch antenna design is given below:

The calculations of the dimensions of the antenna have been illustrated in the following text. These dimensions are useful to create the basic model and then optimize to get the desired results. To design an antenna, the following designing parameters have been taken into account as given below:

t - Very thin metallic strip $t << \lambda_0$

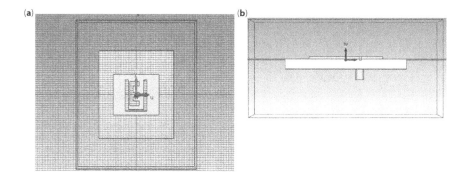

Figure 4.2 Structure of planar microstrip patch antenna.

h- Height of the substrate $h << \lambda_0$, generally $0.003\lambda_0 \leq L \leq 0.05\lambda_0$
L - Length of the patch $\lambda_0 /3 < L < \lambda_0 /2$
W- Width of the patch

Basic length and width calculation formula mention above. The effective length has been taken into account due to the effect of fringing fields. It changes the dimensions of length, so effective length L_e or ΔL is given by:

$$\Delta L = 0.814\, h(\varepsilon_e + 0.3)(W/h + 0.264)/[(\varepsilon_e - 0.258)(W/h + 0.8)] \tag{4.1}$$

$$W = 0.5\lambda_0 / \sqrt[2]{(\varepsilon_e + 1)/2} \tag{4.2}$$

Z_0 is Characteristics impedance,

$$Z_0 = \begin{cases} \dfrac{60}{\sqrt{\varepsilon_e}}\ln\left(\dfrac{8h}{W} + \dfrac{W}{4h}\right) & for\ W\!/_h \leq 1 \\[3mm] \dfrac{120\pi}{\sqrt{\varepsilon_e}\,[W/h + 1.393 + 0.667\ln(W/h + 1.444)]} & for\ W\!/_h \geq 1 \end{cases} \tag{4.3}$$

W/h or W/d ratio can be calculated as:

$$\frac{W}{d} = \begin{cases} \dfrac{8e^A}{e^{2A}-2} & for \; ^W\!/_h < 2 \\[4mm] \dfrac{2}{\pi}\left[B-1-\ln(2B-1)+\dfrac{\varepsilon_r-1}{2\varepsilon_r}\left\{\ln(B-1)+0.39-\dfrac{0.61}{\varepsilon_r}\right\}\right] & for \; ^W\!/_h > 2 \end{cases}$$

$$(4.4)$$

Where;

$$A = \frac{Z_0}{60}\sqrt{\frac{\varepsilon_r+1}{2}} + \frac{\varepsilon_r-1}{\varepsilon_r+1}\left(0.23+\frac{0.11}{\varepsilon_r}\right)$$

$$(4.5)$$

$$B = \frac{377\pi}{2Z_0\sqrt{\varepsilon_r}}$$

Effective length $L_{eff} = L + 2\Delta L$ (4.6)

Ground dimensions (L_g, W_g) - Infinite grounds planes are only used for transmission line models hence we need to consider the finite ground plane for practical designing. It is observed that the result obtained for the infinite ground is similar to considering finite ground length which equals six times the patch length [1,2,13]. Dimension for the ground plane is given by:

$$L_g = 6h + L, \; W_g = 6h + W$$ (4.7)

To operate this model in the fundamental mode TM_{10}, length of patch must take little less than $\lambda/2$, where λ is equal to $\lambda_0/\sqrt{\varepsilon_{reff}}$, where λ_0 = Free space wavelength.

Resonating frequency $(f_r)_{101}$, Dominant mode, L > W when No fringing field.

$$(f_r)_{101} = \frac{1}{2L\sqrt{\mu_0\varepsilon_0\varepsilon_r}}$$ (4.8)

Resonating frequency$(f_r)_{010}$, Dominant mode, L > W with the fringing field.

$$(f_r)_{010} = \frac{1}{2L_{eff}\sqrt{\mu_0\varepsilon_0\varepsilon_{reff}}} \tag{4.9}$$

4.4.3   Calculated and Optimized Value of Antenna

The simulation parameter S_{11} provides the information of the resonating frequency or working frequency range as specified. The optimization of the slotted antenna is necessary for the simulation process to make sure that the resulted parameters are optimum for the desired frequency band. Coordinate for the coaxial feeding (X_f, Y_f) are $(5.4742, 6.895)$. The dimensions of the first patch antenna calculated by patch designing equations are shown in Table 4.1.

The dimensions of the Conformal MSA are the same as Planar MSA and the planar MSA are conform on a cylindrical surface of radius 30 mm. Figure 4.3 represents the slotted antenna on the cylindrical surface and scattering parameter S_{11} shown in Figure 4.4 which is offering resonating frequency on 4.1 and 6.1 GHz.

4.5   Wearable Conformal Antenna

To design a flexible antenna requires thin substrates to accommodate the flexing. Consequently, the radiation characteristics may be degraded due

Table 4.1 Dimension parameter of the patch.

Parameters	Calculated value (mm)	Optimized value (mm)
L_G	38.97	40.53
L_P	19.92	18.48
L_{S1} (U slot)	2	2
W_G	44.83	46.63
W_P	25.78	28.58
W_{S1} (U slot)	14.79	13.79
h	3.175	3.475
L_{s2} (E slot), W_{s2} (E slot)	11, 4	11, 4

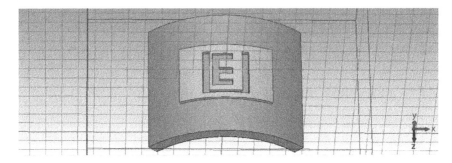

Figure 4.3 Structure of Conformal MSA having E- and U-shape slot.

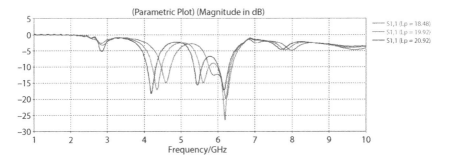

Figure 4.4 Return loss (S11) of the microstrip patch antenna for varying Lp.

to the thinness of the material. Such an antenna system can be developed to operate under such conditions; either the antenna parameter of the curved surface is acceptable for the required application or it has to compensate for the degraded antenna performance using the compensating circuit/components. Thus, implementation of the conformal array may be the solution for different application as civil or military aircraft, vehicles, medical and wearable application and due to the lightweight of such antennas providing the basis for cost-effective mass production.

The field of wearable and flexible technology is wonderfully attractive with its boundless developments in the field of Rf designing, materials science, microwave designing, fabrication and packaging. Wearable and flexible technology, which are lightweight, portable, bendable, reconfigurable, rollable and potentially foldable, would substantially expand the applications of modern electronic devices. Many literature surveys tend to reveal the different considerations for developing proposed antenna using cotton or synthetic cloth materials, CNT, Flexible and semi-flexible substrates and also illustrates wearable antenna effect on the human body and other biomedical application [45], RF- tracking, public safety, in various navigation, telemedicine, and defense purpose [18].

This is the fastest-growing field in the present scenario. Market statistics state that the universal revenue of wearable technology is predicted to rise from about 45 billion dollars in 2016 to around 300 billion dollars in 2028 because it has many advantages, such as as being light in weight, low manufacturing cost, energy proficient, further to reduced fabrication complexity, and the accessibility of less expensive flexible/semi-flexible substrates. Moreover, current developments in miniaturization of the antenna, flexible photo-voltaic, flexible OLEDs, printable energy storage, and green electronic modules have paved the way for the achievements of this technology. Applications of flexible and wearable wireless systems include but are not limited to the fields of health care, personal communication, military communication, industry, entertainment, and aeronautics. The objective of this chapter is to provide all-inclusive guidance to various methodologies and technologies and show how they are applied in the realization of the conformal and flexible antenna.

4.5.1 Wearable Technology

An Equipment or electronic gazette that can comfortably be worn on the living body is called a wearable device. Wearable technology is a developing multidisciplinary field. Knowledge in bioengineering, electrical engineering, software engineering, and mechanical engineering is needed to design and develop wearable communication and medical system. Wearable medical systems and sensors are used to measure and monitor the physiological parameters of the human body. Biomedical systems in the vicinity of the human body may be wired or wireless. Many physiological parameters may be analyzed using wearable medical systems and sensors. Medical systems and sensors for Wearable applications, which are used to measure body temperature, heartbeat, gastro purpose, blood pressure, sweat rate, and any other physiological parameters by wearing the medical device [19, 20]. Wearable technology may provide scanning and sensing features that are not offered by mobile phones or laptop computers. It usually has communication capabilities and users may have access to information in real time. Several wireless technologies are used to handle data collection and processing by medical systems. The collected data may be stored or transmitted to a medical center to analyze the collected data. Wearable devices collect raw data that has to feed to a database or software for analysis. This analysis normally may result in a reaction that might lead an attentive physician to contact a patient who was showing some abnormal signs. However, a similar message may be sent to a person who achieves a fitness goal. Examples of wearable devices include headbands, smart wristbands, belts, watches, glasses, contact lenses, e-textiles, smart fabrics, jewelry, bracelets, and hearing aid devices [18].

Usually, any communication systems comprise a transmitting part, a receiving part, a data processing unit, and wearable antennas. Wearable technology may influence the fields of transportation, healthcare and medicine, fitness, aging, disabilities, education, finance, gaming, media, entertainment, and music. Wearable devices will in the next decade be an important part of individuals' daily lives.

4.5.2   Wearable Devices for Medical Systems

The main aim of wearable systems in medicine is to enhance disease avoidance. A person can be aware of his private health and handle an emerging situation by using the wearable electronic gazette. These system devices measured/recorded medical data continuously and connected for the analysis of the data with a large number of medical centers through which patients may receive better-quality low-cost medical services.

The technical consideration for the human body which affects the human tissue is SAR. Because it is in or in proximity to the human body. The design consideration of such antenna, in which high dielectric constant and power loss produces an unfavorable effect on input impedance and efficiency of an antenna. This power loss in the human body can be characterized by the body-worn efficiency of the device that is given as the ratio of total power that is radiated while the antenna is worn on the body to the total radiated power in free space [21, 47]. Microwave radiation affects the human body while a dielectric material is heated by the electromagnetic waves at higher frequencies [19, 22, 46]. SAR is the factor that is used to quantify the amount of energy the human body tissues absorb. The proposed model should also satisfy the standard SAR limits established by the Federal Communications Commission.

4.5.3   Wearable Medical Devices Applications

Wearable medical systems can be used to:

a)   Help the monitoring of hospital activities
b)   Assist diabetes patients

c) Assist asthma patients
d) Help solve sleep disorders
e) Help solve obesity problems
f) Help solve cardiovascular diseases
g) Assist epilepsy patients
h) Help in the treatment of Alzheimer's disease patients
i) Help gather data for clinical research trials and academic research studies

Several physiological parameters can be measured using wearable medical systems and sensors. Some of these physiological data are presented in this section [9].

4.5.4   Measurement of Human Body Temperature

The temperature of a healthy person ranges between 35°C and 38°C. Temperatures below or above this range may indicate that the person is sick. Temperatures above 40°C may cause death. A person's body temperature may be transmitted to a medical center and if needed the doctor may contact the patient to give further assistance.

4.5.5   Measurement of Blood Pressure

Usually, blood pressure and heartbeat are measured in the same set of measurements. A blood pressure measurement indicates the arterial pressure of the blood circulating in the human body. Some of the causes of changes in blood pressure may be stress and being overweight. The blood pressure of a healthy person is around 80 by 120, where the systole is 120 and the diastole is 80. Changes of 10% above or below these values are a matter of concern and should be examined. The blood pressure and heartbeat may be transmitted to a medical center and if needed the doctor may contact the patient to give further assistance.

4.5.6   Measurement of Heart Rate

Measurement of the heart rate is one of the most important tests when examining the health of a patient. A change in heart rate will change the blood pressure and the amount of blood delivered to all parts of the body. The heart rate of a healthy person is 72 beats per minute. Changes in heartbeat may cause several kinds of cardiovascular disease. Traditionally heart rate is measured using a stethoscope. However, this is a manual test and is not so accurate. To measure and analyze the heartbeat a wearable medical device may be connected to a patient's chest. Medical devices that measure heartbeat can be wired or wireless [23].

4.5.7   Measurement of Respiration Rate

Measurement of respiration rate indicates if a person is breathing normally and if the patient is healthy. Elderly and overweight people have difficulty breathing normally. Wearable medical devices are used to measure a person's respiration rate. A wired medical device used to measure respiration rate may cause uneasiness to the patient and cause an error in measurements of respiration rate. It is better to use a wireless medical device to measure respiration rate. The measured respiration rate may be transmitted to a medical center and if needed the doctor may contact the patient for further assistance.

4.5.8   Measurement of Sweat Rate

Glucose is the primary energy source of human beings. Glucose is supplied to the human body usually as a monosaccharide sugar that provides energy to the human body. When a person does extensive physical activity, glucose comes out of the skin as sweat. A wearable medical device can be used to monitor and measure the sweat rate of a person during extensive physical activity. A wearable medical device can be attached to the person's clothes in proximity to the skin to monitor and measure the sweat rate. This device can also be used to measure sweat Ph, which is important in the diagnosis of diseases. Water vapor evaporated from the skin is absorbed in the medical device to determine the sweat Ph. If the amount of sweat coming out of the body is too high, the body may dehydrate. Dehydration causes tiredness and fatigue. Measurements of sweat rate and Ph may be used to monitor the physical activity of a person.

4.5.9   Measurement of Human Gait

The movement of human limbs is called human gait. Different limb movement patterns, characterize the various gait patterns such as force, velocity, kinetic energy and potential energy cycles, and also ground contact patterns. Walking, jogging, skipping, and sprinting are defined as natural human gaits. Gait analysis is a helpful and fundamental research tool to characterize human locomotion. Wearable devices may be attached to various parts of the body to measure and examine human gait. The analysis of the human gait is performed by a recording of the movement signals. Recording and estimation of the temporal features of the gait are performed by the pressure sensors and the wearable accelerometers that are placed inside the footwear. The ambulatory gait analysis results may determine the type of treatment for the patient [23, 24].

Gait analysis has been proved as a trustworthy diagnostic technique for diseases with neurological disorders such as Parkinson's disease and stroke. Parkinson's disease affects the human gait movements as it slows the movement of the limbs. Analysis of such cases has been performed effectively using the wearable devices in health monitoring systems (shown in Figure 4.5). Body movement abnormalities may be predicted for health problems or the progression of neurodegenerative diseases. Due to such abnormalities, people may fall suddenly which can become a serious threat to the

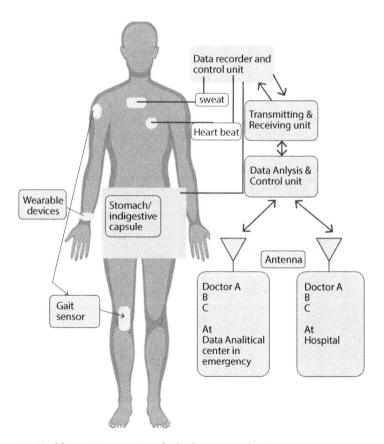

Figure 4.5 Health monitoring system for body-worn application.

health of elderly people. Gait analysis using wearable devices has been used to analyze and predict falls among elderly patients [18, 23-26].

4.6   Textile and Cloth Fabric Wearable Antennas

Recent developments in wireless communication technology have paved the way for the growth and development of wearable textile antennas built using a diverse variety of fabrics. It is different from a conventional antenna and can be fabricated on a part of the clothes and body. It is similar to a printed antenna, a textile or cloth-based material is used as a substrate and cupper as a radiating element [9, 27, 28]. The wearable antennas have been developed for applications where one is working while wearing it on the body (shown in Figure 4.6). The present literature proposes that these antennae may be integrated on textiles such as polyester, polyethylene, polyfloncuflon (tm), Polystyrene, Polyamide, nylon, cotton, foam, conducting paint, insulated wire, liquid crystal polymer, etc. [12, 49-52].

4.6.1   Specific Absorption Rate (SAR)

The specific absorption rate is used to measure the EM radiation that is emitted by antennas/mobile towers/cell phones. It is measured in watt per

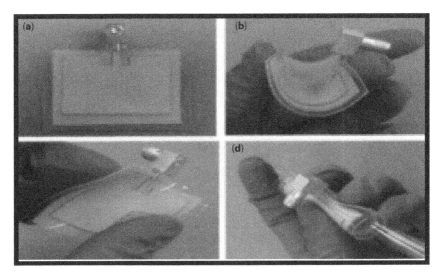

Figure 4.6 Flexible wearable antenna [28].

kilogram of tissue (W/kg) which is power absorbed/kg of tissue. This is typically averaged over a small section volume (1 gm or 10 gm of tissue) [11, 29]. The allowable maximum limit for mobile phones is set by the Federal Communications Commission (FCC) that is 1.6 W/kg of EM radiation that accounts for just 6 minutes per day. Elsewhere, a person may use their cell phone within the safety margin limits of 18 to 24 minutes per day.

Calculation of SAR can be estimated using the given formula:

$$\text{Specific Absorption Rate} = \sigma\, E/\rho\ (\text{W/kg}) \qquad (4.10)$$

where E is the RMS electric field, σ is the electrical conductivity, and ρ is the sample density

4.6.2 Interaction with Human Body

In Section 4.3, wearable antenna technology and application in the medical system have been discussed. Section 4.2 describes the antenna designing parameters considering the interaction of the human body with the medical systems for wearable applications. The interaction of the wearable antenna with the human body can be analyzed as follows. It is possible to achieve using an on-line platform which determines tissues characteristics which may have 1-8 layers for all high-frequency range [30].

A 6-layer arm model has been proposed that work includes 1st skin tissue, blood tissue, fat tissue, nerves tissue, muscle tissue, and bone tissue. SAR simulations proposed using HFSS for an arm as it is a complex structure with different types of tissues [29, 31]. As the tissue characteristics are different at frequencies, here following characteristics for 0.406GHz has been discussed. The most important body organ is the brain; several research and project- related studies on EM radiation effects of high-frequency for the head tissue have been done previously. Analysis of density, thickness and electrical parameters of the human arm tissue are illustrated in Tables 4.2 and 4.3 [31].

4.6.3 Wearable Devices Tracking and Monitoring Doctors

Each patient can have a wearable device attached to the body. The wearable device is connected to several sensors where every particular sensor performs its specific job. For example, a sensor node may detect heart rate and body temperature while the other detects blood pressure. Medical

Table 4.2 Density and thickness of arm tissues mass [31].

Tissue	Mass density [kg/m$_3$]	Thickness [mm]
Bone	1908	17
Blood	1050	4
Nerve	1075	1
Muscle	1090	15
Fat	911	1.5
Skin	1109	1.5

Table 4.3 Arm tissues parameters at distress signal frequency (0.406GHz) [31].

Tissue	Conductivity [S/m]	Relative permittivity	Loss tangent
Bone	0.029385	5.6689	0.22947
Blood	1.3516	64.115	0.93324
Nerve	0.44832	35.339	0.56161
Muscle	0.79787	57.079	0.6188
Fat	0.041216	5.5773	0.32715
Skin	0.69064	46.649	0.65541

professionals may also carry a wearable device, which allows other hospital personnel to locate them within the hospital.

4.6.4   Wireless Body Area Networks (WBANs)

In recent years, many of the textile substrates came into the picture for the fabrication of wearable antennas. Based on substrate properties, a comparative analysis for different substrates has been proposed for industrial, scientific, and medical (ISM) applications and miniaturization of wireless devices [32]. The main objective of WBANs is to provide continuous bio-medical data for a person that may record blood pressure, heartbeat rate, electrocardiograms, electro-dermal activity, body temperature, and several other parameters of the body. For example, gyroscopic sensors can be used to detect different types of body movements and muscular activities [18, 20, 24]. WBANs have a wide range of applications for communication, wearable system [7].

WBAN is in high demand for various applications in different fields such as sports, military, healthcare monitoring [8, 33]. Detailed research has been carried out for synthetic and non-synthetic material used as substrate materials for wearable antenna or body-worn applications. Several studies of cotton, foam, polymers, and other fabrics have been projected which are used as flexible substrate materials, which would be integrated on the human body and clothing [9, 28, 33]. The substrate used for integration is required to be semi-flexible/flexible, robust, and highly resistive to environmental features; they are moisture, humidity, vapor ionization, thermal effects, which is appropriate for wearable system applications. It has a benefit for the users of the wearable antenna as it is integrated on the part of clothing, and used to transmit and receive electromagnetic signals through the textile fabric. Such wearable antennas perform a vibrant role in medicine, health monitoring, physical training, RFID, navigation, tracking, and military applications [34].

4.7   Design of Liquid Crystalline Polymer (LCP) Based Wearable Antenna

The focus of this chapter is to propose a theory for Conformal wearable antenna designing, implementation and challenges with present procedures, design methodologies, software tools and framed works of the wearable conformal antenna using a polymer substrate having a dielectric constant of 2.9.

A polymer substrate microstrip antenna is proposed for the purpose to verify the radiation directivity and other antenna parameters for millimeter-wave application at 38GHz. Consequently, the selected substrate is a Rogers 3850- an LCP substrate material that has a dielectric constant εr of 2.9, loss tangent $\tan\delta$ of 0.0025 and thickness t = 1.1 mm with ultra-thin thickness. A slot structure is etched from the ground plane of the proposed antenna, which is known as Defected Ground Structure (DGS). This is a developing technique used for the enhancement of various antenna parameters as narrow bandwidth, cross-polarization, low gain, etc. W. Y. Tam presented integral equations to include the spectral domain green's function and unknown slot electric field, where the author has illustrated the numerical solution using the method of moments [35]. Wong Kin-Lu proposed a line feed printed wide-slot antenna with a tuning stub used to enhance the bandwidth [17].

4.7.1   Dimensions of the Proposed Model

In this planar antenna, a guard line is incorporated to both sides of the patch to reduce surface current and observe the effect of the guard line on its resonating frequencies. The dimensions of the proposed model are given in Table 4.4.

4.7.2   Slot Loaded Ground: (Defective Ground Structure - DGS)

DGS slot is etched from the ground that may periodic or non-periodic cascaded configuration or any other shapes, such planar transmission lines that interferes with the distribution of the shield current in the ground plane that causes the ground defect. This current interference can alter a transmission line's characteristics, such as line inductance and capacitance. In a nutshell, any defect etched into the microstrip's ground plane will lead to increased productive capacitance and inductance [36]. Different profiles of the slot are in the shape of E, H, U, etc. The easy way to achieve sufficient bandwidth by appropriate slot dimensions is to use the slot. A probe feeding antenna proposed with impedance bandwidth of 103.6%, return loss of -30 dB for C- band [37, 38]. A slot etched from the ground to improve the BW [39]. The DGS slot with dimension is shown in Figure 4.7.

Table 4.4 Dimensions of the proposed planar antenna.

Sr. no.	Component	Patch dimension	Slot dimensions
1	Patch length & width	9mm × 5.69mm	5 segment cylinder R = 0.5mm, Height = 1.1mm
2	guard line1 and 2	5.65mm × 1.2mm	Cylinder R = 0.7mm, Height = 1.1mm
3	Guard line 3	1.2mm × 7mm	Box = 0.2 × 4 × 1.1mm
4	Ground length & width	57mm × 30.6mm	

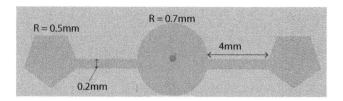

Figure 4.7 A defective ground slot with polymer substrate ε_r of 2.9.

The proposed planar and wearable antenna modal shown in Figure 4.8 using HFSS and CST simulation tool.

The antenna resonating between 7.5-10.5GHz for the different textile substrates and offering return losses for all substrates near -40dB. It is also observed that it has a little lower directivity in comparison with other textile substrates. In general, Polyester materials are less. The height of dielectric material is a bit critical to select for the conformal structure manufacture because the flexibility of the material is easy to fold/bend for bit thin materials. Figure 4.9 presents the scattering parameter analysis as a reflection coefficient for different substrates.

4.7.3   Radiation Characteristics

The radiation result analysis state that the radiation pattern is significantly affected in terms of angular width and radiation direction for the different substrates as Polyester, Polyethylene, Polyfloncuflon (tm), Polystyrene, Polyamide. In addition of resonant frequency is also slightly affected by

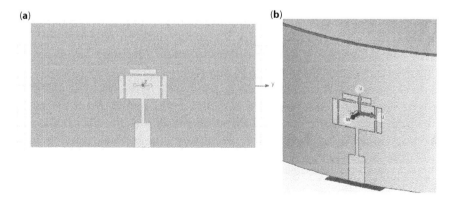

Figure 4.8 A defective planar and conformal wearable antenna with a polymer substrate.

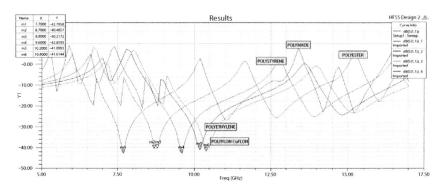

Figure 4.9 Scattering parameter S_{11} for different textile substrate.

little change in curvature of the patch (shown in Figure 4.10). The radiation-pattern in the direction of elevation relies greatly on the radius of the cylinder, but the angle of the azimuth depends much less on the cylinder radius [40–43].

4.8   Result Discussion and Analysis

This section provides a detailed analysis of simulated in terms of scattering parameter and radiation characteristics parameter of the proposed antenna. Theoretical aspects linking to the results will be discussed and commented on Design and Analysis of a slotted conformal antenna discussed in section 4.2. Furthermore, a DGS conformal antenna has been illustrated for the wearable application using various textile substrates. Due to which antenna is on a single resonating frequency from 7.5 GHz to 10.5 GHz x-Band application for 5 substrates with an acceptable directivity and gain range up to 2-4.5 dBi for substrate Polyester, Polyethylene, Polyfloncuflon (tm), Polystyrene, Polyamide.

It is also shown that the impedance bandwidth of the wearable antenna can be significantly improved by using DGS configuration. Since the gain-bandwidth product remains constant, an attempt has been made to enhance the patch antenna bandwidth while maintaining the optimal radiation pattern. The return loss increases in the E plane when the antenna architecture is closely spaced. It is illustrated that on increasing the spacing of the array, the antenna-gain is greatly reduced. Also, by optimizing the guard-line spacing on both patches, the effects of patch surface waves and mutual coupling may be reduced. For thin and thick dielectric substrates with low permittivity, the process has been successfully analyzed. It has been observed that the antenna built for the given resonant frequency corresponds to the patch dimension with great accuracy. The empirical relationships derived for curvature could also be used to determine the antenna designing parameters for a microstrip antenna subject to the limits of substrate layer thickness (0.58 mm-1.54 mm), scattering parameter, resonant frequency, return loss, power radiated, directivity and gain. To overcome the threat under conformality, wearable antenna systems may need to be reconfigurable and functional to overcome intentional and unintentional electromagnetic disturbances.

With the center-frequency set as 8.357 GHz, the proposed models achieve frequency agility spanning from 15.5 percent to 24 percent. For applications where large (azimuthal) angular coverage is needed, antennas mounted on single curved cylindrical surfaces are an essential class of conformal arrays. In radar and communication networks, these antenna styles may be used.

4.9   Challenges and Future Needs

Remarkable advances for the proposed work softness, flexibility, and ability of integration of circuits has already been achieved. But the noticeable hurdles are bio-safety, re-usability or wear resistance, shelf life, and its longitudinal performance, which has to be addressed. The objective is to focus on some areas of opportunity for the future prospective study to give a strong contribution towards the advancement, accuracy, broad level commercialization and the adoption of wearable antenna systems.

The antenna should conform to the soft and stretchable dielectric. Low backward power radiation, less signal loss, and low power consumption are the desirable feature of these antennas [48]. It is very difficult to replicate the antenna on HFSS/CST, to simulate it together for the different parts of the human body.

4.10   Conclusion

Slotted wearable conformal antenna configuration has been investigated. A DGS conformal antenna was also demonstrated for wearable use utilizing different textile substrates. Because of which antenna is on a single resonating frequency from 7.5GHz to 10.5 GHz x-Band operation for 5 substrates

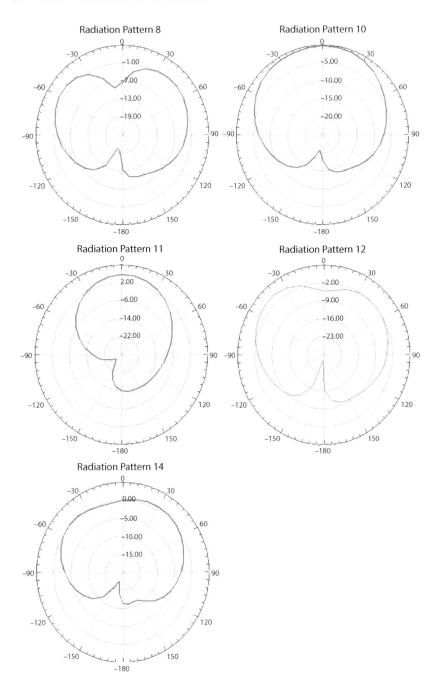

Figure 4.10 Radiation characteristics of the wearable antenna (a) Polyester;
(b) Polyethylene; (c) Polyfloncuflon (tm); (d) Polystyrene; (e) Polyamide substrate.

with appropriate path and scale up to 2-4.5 dBi for polyester, polyethylene, polyfloncuflon (tm), polystyrene, polyimide substrates.

It is also seen that wearable antenna impedance bandwidth can be substantially increased using DGS configuration. As the product of gain and bandwidth always remain constant, an effort has been made to increase patch antenna bandwidth thus maintaining the preferred radiation pattern. In addition, the consequences of surface waves and reciprocal coupling may be reduced by improving guard line width on both patches. The method was effectively analyzed on low-permittivity thin and thick dielectric substrates. The antenna built for a given resonant frequency was found to fit the patch dimension with high precision.

The suggested designs gain frequency agility varying from 15.5% to 24% with core frequency set at 8,357 GHz. Antennas installed on single-curved cylindrical surfaces are an essential class of conformal arrays for applications needing broad (azimuthal) angular coverage. They can be found in radar and networking networks.

References

1. Ahmed Emad S., "Conformal Band-Notch UWB Monopole Antenna On Finite Cylindrical Substrates" *ETASR - Engineering, Technology & Applied Science Research* Vol. 3, No. 3, pp. 440-445, 2013.
2. Balanis C. A., *Antenna Theory: Analysis Design*, Third Edition, ISBN 0-471-66782-X 2005, New York, John Wiley & Sons, 1997.
3. Biswas Diptiman and Ramachandra V., "A Novel Approach to Design Conformal Frustum Wrap Around Antenna", *9th International Radar Symposium India, IRSI* - 13, pp. 10-14, 2013.
4. C. Du, S. Zhong, L. Yao, and Y. Qiu, "Textile microstrip array antenna on three-dimensional orthogonal woven composite," in *Proceedings of the 4th European Conference on Antennas and Propagation (EuCAP '10)*, pp. 1–2, April 2010.
5. C. Hertleer, A. Tronquo, H. Rogier, and L. Van Langenhove, "The use of textile materials to design wearable microstrip patch antennas," *Textile Research Journal*, vol. 78, no. 8, pp. 651–658, 2008.
6. Hertleer C., Laere A.V., Rogier H., Langenhove L.V. "Influence of Relative Humidity on Textile Antenna Performance" *Text. Res. J.* 80:177–183, 2009.
7. S. Bhavani, T. Shanmuganantham "Analysis of Different Substrate Material on Wearable Antenna for ISM Band Applications" conference *Advances in Communication Systems and Networks* pp. 753-762, June 2020.

8. E. F. Sundarsingh *et al.*, "Polygon-Shaped Slotted Dual-Band Antenna for Wearable Applications," *IEEE Antennas Wirel. Propag. Lett.*, vol. 13, pp. 611-614, 2014.

9. M. El Atrash, K. Bassem, and M. A. Abdalla, "A Compact Dual-Band Flexible CPW-fed Antenna for Wearable Applications," *2017 IEEE Int. Symp. Antennas Propag. Usn. Natl. Radio Sci. Meet.*, pp. 2463–2464, 2017.

10. S. R. Zahran, M. A. Abdalla, and A. Gaafar, "Time Domain Analysis for Foldable Thin UWB Monopole Antenna," *AEU-International Journal of Electronics and Communications*, vol. 83, pp. 253-262, 2018.

11. Titti Kellomaki, "Effect of Human Body on Single Layer Wearable Antenna", Tampere University of Technology publication 1025.

12. Ankita and Brajlata Chauhan "A Review of Textile and Cloth Fabric Wearable Antennas" *International Journal of Computer Applications* (0975–8887) Volume 116, No. 17, April 2015.

13. Brill Yishai, IAI –MLM "Construction and Calculation of Conformal Antennas" CST 2008 vs. CST 2009 IAI Ltd, 2009.

14. Gupta Samir Dev *et al.* "Design and Performance Analysis of Cylindrical Microstrip Antenna and Array using Conformal Mapping Technique" *International Journal of Communication Engineering Applications IJCEA*, Vol. 02, Issue 03, pp. 166-180, 2011.

15. Josefsson Lars and Persson Patrik, *Conformal Array Antenna Theory and Design*, IEEE Antennas and Propagation Society, Sponsor, Hoboken, NJ: John Wiley and Sons, Ltd, 2006.

16. Jong Jin Baek, Se Woong Kim and Youn Tae Kim "Camera-Integrable Wide-Bandwidth Antenna for Capsule Endoscope" *Sensors* 2020, *20*(1), 232; , 31st Dec. 2019.

17. Sze Jia-yi and Wong Kin-Lu [2001], "Bandwidth enhancement of microstrip line fed printed wide slot antenna." In: *IEEE Transactions on Antennas and Propagation*, Vol.49, Issue 7, pp. 1020 – 1024.

18. T. Gao, D. Greenspan, M. Welsh, R. R. Juang, and A. Alm. "Vital signs monitoring and patient tracking over a wireless network", *Proceedings of the 27th Annual International Conference of the IEEE EMBS, Shanghai, China, 1–4 September 2005*; pp. 102–105.

19. A. T. Barth, M. A. Hanson, H. C. Powell, Jr. and J. Lach, Tempo 3.1: A body area sensor network platform for continuous movement assessment. *Proceedings of the 6th International Workshop on Wearable and Implantable Body Sensor Networks, Berkeley, CA, June 2009*, pp. 71–76.

20. M. Gietzelt, K. H. Wolf, M. Marschollek and R. Haux, "Automatic self-calibration of body worn triaxial-accelerometers for application in healthcare" *Proceedings of the Second International Conference on Pervasive Computing Technologies for Healthcare, Tampere, Finland, January 2008*, pp. 177–180.

21. Albert Sabban "Novel Wearable Antennas for Communication and Medical Systems," CRC Press, Taylor & Francis Group, Boca Raton, FL 33487-2742, 2018.

22. Y. Rahmat-Samii, "Wearable and Implantable Antennas in Body-Centric Communications". *Second European Conference on Antennas and Propagation* 2007. EuCAP 2007.

23. S. J.M. Bamberg, A. Y. Benbasat, D. M. Scarborough, D. E. Krebs and J. A. Paradiso, Gait analysis using a shoe-integrated wireless sensor system. *IEEE Transactions on Information Technology in Biomedicine*, vol. 12, 413–423, 2008.

24. A. Purwar, D. U. Jeong and W. Y. Chung, "Activity monitoring from realtime triaxial accelerometer data using sensor network" *Proceedings of International Conference on Control, Automation and Systems*, Hong Kong, 21–23 March 2007, pp. 2402–2406.

25. J.H. Choi, J. Cho, J.H. Park, J.M. Eun and M.S. Kim, An efficient gait phase detection device based on magnetic sensor array. *Proceedings of the 4th Kuala Lumpur International Conference on Biomedical Engineering, Kuala Lumpur, Malaysia, 25–28*; 21, pp. 778–781, June, 2008.

26. J. Hidler, Robotic assessment of walking in individuals with gait disorders. *Proceedings of the 26th Annual International Conference of the IEEE Engineering in Medicine and Biology Society, San Francisco, CA*, 1–5; 7, pp. 4829–4831, September 2004.

27. R. B. V. B. Simorangkir, Y. Yang, L. Matekovits, and K. P. Esselle,"Dual-Band Dual-Mode Textile Antenna on PDMS Substrate for Body- Centric Communications," *IEEE Antennas Wirel. Propag. Lett.*, vol. 16, pp. 677-680, 2017.

28. S. Yan, P. J. Soh, and G. A. E. Vandenbosch, "Compact All-Textile Dual-Band Antenna Loaded With Metamaterial-Inspired Structure," *IEEE Antennas Wirel. Propag. Lett.*, vol. 14, pp. 1486-1489, 2015.

29. Francesc Soler "Radiation Effects of Wearable Antenna In Human Body Tissues" Heather Song University of Colorado Springs, 2014.

30. IFAC, "Dielectric properties of body tissues", http://niremf.ifac.cnr.it/tissprop/.

31. Rifaah Zaki Alkhamis, "Global positioning system (GPS) and distress signal frequency wrist wearable dual-band antenna", UCCS, retrieved July 2013.

32. F. Declercq, I. Couckuyt, H. Rogier, and T. Dhaene, "Complex permittivity characterization of textile materials by means of surrogate modelling," in *Proceedings of the IEEE International Symposium Antennas and Propagation and CNCUSNC/URSI Radio Science Meeting*, July 2010.

33. S. Velan *et al.*, "Dual-Band EBG Integrated Monopole Antenna Deploying Fractal Geometry for Wearable Applications," *IEEE Antennas Wirel. Propag. Lett.*, vol. 14, pp. 249-252, 2015.

34. Mohamed Kadry and Mohamed El Atrash, "Design of an Ultra-thin Compact Flexible Dual-Band Antenna for Wearable Applications"

AP-S 2018, IEEE Conference Paper· August 2018 DOI: 10.1109/APUSNCURSINRSM.2018.8609247

35. W.Y, Tam *et al.*, 'Microstripline-fed cylindrical slot antennas' *IEEE Transactions on Antennas and Propagation*, Volume: 46, Issue: 10, pp. 1587 – 1589, 1998.

36. Dian Wang, *et al.*, "Small Patch Antennas Incorporated With a Substrate Integrated Irregular Ground" *IEEE Transactions on Antennas and Propagation*, Vol. 60, No. 7, pp. 3096-3103, 2012.

37. Fan Yang *et al.*, "Wide band E- shaped patch antennas for wireless communications" *IEEE transactions on antennas and propagation*, Vol. 49, No. 7. pp. 1095-1099, 2001.

38. Kahrizi Masoud and Sarkar Tapan K., "Analysis of a wide radiating slot in the ground plane of a microstrip line" *IEEE Transactions on Microwave Theory and Techniques*, Vol. 41, Issue 1, pp. 29–37, 1993.

39. Sze Jia-yi and Wong Kin-Lu, "Bandwidth enhancement of microstrip line fed printed wide slot antenna." In: *IEEE Transactions on Antennas and Propagation*, Vol. 49, Issue 7, pp. 1020–1024, 2011.

40. Alomainy, A.; Hao, Y.; Owadally, A.; Parini, C.; Nechayev, Y.; Constantinou, C.C.; Hall, P. Statistical Analysis and Performance Evaluation for On-Body Radio Propagation with Microstrip Patch Antennas. *IEEE Trans. Antennas Propag.* 2007, 55, 245–248.

41. Yan, S.; Soh, P.J.; VandenBosch, G.A.E. Wearable Dual-Band Magneto-Electric Dipole Antenna for WBAN/WLAN Applications. *IEEE Trans. Antennas Propag.* 2015, 63, 1.

42. G.-Y. Lee, D. Psychoudakis, C.-C. Chen, and J. L. Volakis, "Omnidirectional vest-mounted body-worn antenna system for UHF operation," *IEEE Antennas and Wireless Propagation Letters*, vol. 10, pp. 581–583, 2011.

43. C. Hertleer, H. Rogier, L. Vallozzi, and L. Van Langenhove, "A textile antenna for off-body communication integrated into protective clothing for firefighters," *IEEE Transactions on Antennas and Propagation*, vol. 57, no. 4, pp. 919–925, 2009.

44. L. Vallozzi and H. Rogier, "Radiation characteristics of a textile antenna designed for apparel application," in *Proceedings of the IEEE Symposium for Space Applications of Wireless and RFID (SWIRF '07)*, May 2007.

45. Li, Y.; Guo, Y.X.; Xiao, S. Orientation insensitive antenna with polarization diversity for wireless capsule endoscope system. *IEEE Trans. Antennas Propag.* 65, 3738–3743, 2017.

46. Linheng He, Kechun Wen, *et al.*, "Advanced materials for flexible electrochemical energy storage devices", Journal of Material Research, published online by Cambridge University Press: 26 July 2018.

47. S. Agneessens, and H. Rogier, "Compact Half Diamond Dual-Band Textile HMSIW On-Body Antenna," *IEEE Transactions on Antennas and Propagation*, vol. 62, no. 5, pp. 2374-2381, 2014.

48. Yan, S.; Volski, V.; VandenBosch, G.A.E. Compact Dual-Band Textile PIFA for 433-MHz/2.4-GHz ISM Bands. *IEEE Antennas Wirel. Propag. Lett.* 2017, 16, 2436–2439.

49. Moro, R.; Agneessens, S.; Rogier, H.; Dierck, A.; Bozzi, M. Textile Microwave Components in Substrate Integrated Waveguide Technology. *IEEE Trans. Microw. Theory Tech.* 2015, 63, 422–432.

50. Maria Lucia *et al.* [2012], "High-Gain Textile Antenna Array System for Off-Body Communication" *International Journal of Antennas and Propagation,* Volume 12, pp. 52-57.

51. N. H. M. Rais, P. J. Soh, M. F. A. Malek, and G. A. E. Vandenbosch, "Dual-Band Suspended-Plate Wearable Textile Antenna," *IEEE Antennas Wirel. Propag. Lett.,* vol. 12, pp. 583-586, 2013.

52. Mariam El Gharbi *et al.,* "A Review of Flexible Wearable Antenna Sensors: Design, Fabrication Methods, and Applications" Material Aug. 2020.

5

Design and Analysis of On-Body Wearable Antenna with AMC Backing for ISM Band Applications

B Prudhvi Nadh[1*] and B T P Madhav[2]

[1]*Dhanekula Institute of Engineering and Technology, Vijayawada, India*
[2]*Koneru Lakshmaiah Education Foundation, Vijayawada, India*

Abstract

Current research has mostly concentrated on the wearable devices to track the Human activity monitoring system, improvement of the health care facility and communication between the users aimed for the wellbeing of society. The growing attention given to wearable device technologies is attracting Smart clothing and facilitating communication between people and wireless devices. In this process, wearable antennas are playing an important role in the communication between clothes and the surrounding environment. In most cases the antenna operates in contact with human bodies, which is a hostile environment for the electromagnetic waves for propagation. The problem can be resolved by adopting an appropriate design with reduced backpropagation using AMC structure, which will allow propagation of waves without interfering with the human body and reduce the Specific Absorption Rates.

This chapter concentrates on the design aspects of Wearable Antenna for ISM (Industrial Scientific Medical), WBAN and other Medical applications. The designed antenna with the placement of AMC (Artificial Magnetic Conductor) as backing to improve the gain and communication capacity is proposed. The gain enhanced AMC backed flexible antenna is studied for its performance characteristics with different locations of the phantom. The study is carried out by measuring different parameters of the antenna. The designed antenna is compared with prototype antenna measurement results, and real-time applications are analyzed.

Keywords: Specific absorption rate (SAR), artificial magnetic conductor (AMC), industrial scientific medical (ISM)

*Corresponding author: prudhvibadugu@kluniversity.in

Prashant Ranjan, Dharmendra Kumar Jhariya, Manoj Gupta, Krishna Kumar, and Pradeep Kumar (eds.)
Next-Generation Antennas: Advances and Challenges, (91–104) © 2021 Scrivener Publishing LLC

5.1   Introduction

Wearable technology is one of the powerful tools in medical and monitoring devices in wireless body area networks (WBNS). The parameters measured by wearable devices are blood pleasure, temperature, heart rate and electrodermal activity of healthcare parameters [1-4]. The collected data can be recorded, stored and analyzed by doctors. In wearable technology biomedical antennas usage has increased in this decade. These wearable antennas are worn by people and can connect with external wireless devices [5]. To provide the connectivity antennas should be robust towards environmentally resistive and should be conformal towards the human surfaces [6-8]. These antennas will be placed close to human bodies, due to their high lossy nature distortion of radiation and shift the operating frequency of the antenna. These wearable antennas are designed using dielectric substrates with a minimal amount of mechanical deformation and the influence of weather conditions [9-12].

There are many types of fabric and non-fabric materials that can be used in wearable antenna design [13-15]. Most of the non-textile antennas are made from polymer materials that have nonconductive nature and materials like Kapton polyimide, polyethylene terephthalate (PET) come under this nonconductive fabric. The fabric materials like jeans, cotton and jute are used as substrate materials. In general, all these fabric materials should have high conducting radiating elements with constant thickness and low dielectric constant which helps to store electrical energy. In this chapter, we mainly concentrated on the design of a wearable antenna for on-body applications. The circular ring with a star-shaped antenna is inserted in the design structure to obtain better resonating antenna performance. The designed antenna is placed on the flexible substrate material which can be used for conformal applications when placed on human phantom structures. The designed antenna covering the ISM frequency band is suitable for biomedical applications.

5.2   Design of Star-Shape with AMC Backed Structure

The circular ring antenna with star-shaped antenna is designed on a flexible polyimide antenna and is done having a dielectric constant of 3.5. The antenna has the dimensions of $21 \times 12 \times 0.1$ mm^3 and with a substrate loss tangent of 0.0025. The designed antenna is done in the iteration process starting from the circular ring with partial ground structure. In the second iteration, an antenna is loaded star-shaped structure on the patch presented in Figure 5.1. Further, the antenna is modified with a small circle loaded on the circular ring with a radius of 1 mm. For the lower SAR value and improve antenna gain values is loaded with AMC backed structure. The dimensions of the AMC structure 30 x 30 x 0.1mm³ and also presented in Table 5.1.

The radius of the circular patch is defined by 'R_1' is calculated using [14].

$$R_1 = \frac{F}{\left\{1 + \frac{2h}{\pi\varepsilon_r}\left[\ln\left(\frac{\pi F}{2h}\right) + 1.7726\right]\right\}^{1/2}} \tag{5.1}$$

where 'h' height of the substrate.

The substrate having a permittivity of 'ε_r'

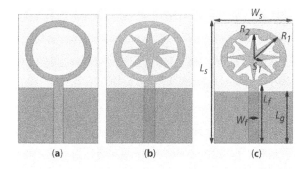

Figure 5.1 Designed antenna by iteration (a) Circular ring (b) Inserted star shape (c) proposed antenna.

Table 5.1 Variables of the designed antenna and AMC.

Variable	Value(mm)	Variable	Value(mm)
R_1	5	L_s	21
R_2	4	W_s	12
L_f	10	S_1	1.4
W_f	1.5	L_g	9
r_2	2	w_u	12
s	2	d	4
r_1	3	L_u	12

5.2.1   Characterization of AMC Unit Cell

Design of AMC structure should have perfect magnetic conductor characteristics [15]. These characteristics with zero reflection phase at the operating frequency are required for AMC. Figure 5.2 gives information about an AMC unit cell designed with substrate permittivity 3.5. The numerical validation of unit cell is done in ANSYS to analyze the reflection phase of the AMC unit cell.

In this article, a 2 × 2 array of the unit cell is arranged which is shown in Figure 5.3. To analyze unit cell is carried by applying the boundary conditions PEC and PMC along x and y-direction. The flower-shaped structure is loaded inside the rectangular ring structure and obtained by applying master-slave boundary conditions.

Figure 5.4 presents the designed antenna with a fabricated model with dimensions indication. The observed changes in S_{11} can be seen in Figure 5.5.

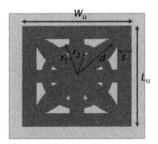

Figure 5.2 Structural view of the unit cell.

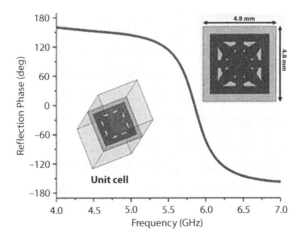

Figure 5.3 Proposed AMC phase diagram of AMC unit cell.

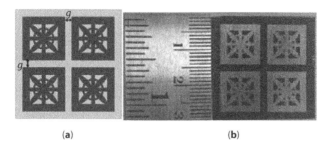

(a)   **(b)**

Figure 5.4 AMC structure (a) 2 × 2 array AMC structure (b) Fabricated AMC.

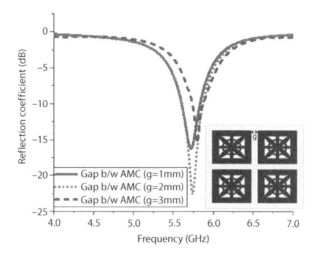

Figure 5.5 Gap variations among the AMC unit cell structure.

If the gap among AMC elements is adjusted to 1 mm antenna covers 5.6-5.83 GHz with minimum S_{11} at 5.72 GHz. If the gap between the elements is increased to 2 mm the operating band ranges from 5.61-5.86 GHz with the resonating frequency of 5.76 with -22.6 dB S_{11}. When the gap between the elements is changed to 3 mm it covers a frequency of 5.7-5.9 GHz. The observation is the shift in the resonating frequency when the gap is increased between elements.

5.3   Discussion of Results of Star-Shaped Antenna with AMC Structure

The designed star-shaped antenna started with the design of the circular antenna operating band of 5.2-7 GHz having a bandwidth of 1.8 GHz with

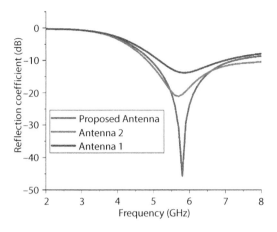

Figure 5.6 Reflection coefficient of proposed antenna iterations.

S_{11} of -13 dB and covering WLAN and ISM band applications. The further antenna is modified with a loading star-shaped structure in the circular ring and the antenna covers the frequency range 4.9-8.4 GHz with a bandwidth of 3.5 GHz and S_{11} of -21dB at 5.7 GHz. Finally, it is loaded with semicircles attached to the circular ring and antenna covers of 5-7.2 GHz with a bandwidth of 2.2 GHz with S_{11} of -45 dB at the resonating frequency of 5.8 GHz and covering ISM frequency applications presented in Figure 5.6.

5.3.1   Bending Analysis of Star-Shaped Antenna with AMC Backed Structure

Bending analysis of star-shaped design is done by keeping AMC backed to star-shaped antenna shown in Figure 5.7. If the bending angle is 30^0

Figure 5.7 Bending analysis of the star-shaped design with AMC backed.

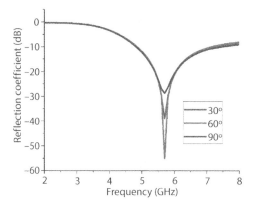

Figure 5.8 Variation of reflection coefficient antenna at different bending conditions.

covers a frequency band of 5.83-5.87 GHz having S_{11} of –13.1 dB. When an angle is increased to 60^0, covers a frequency band of 5.8-5.88 GHz with S_{11} of –17 dB. Similarly, when it is further increased to 120^0 shift in the operating band is observed which covers the frequency band of 5.84-5.88 GHz and the resonating frequency at 5.86 GHz presented in Figure 5.8.

5.4 On-Body Placement Analysis of Proposed Antenna with AMC Structure

5.4.1 Specific Absorption Rate Analysis

The designed antenna for wearable applications, a human phantom structure is shown in Figure 5.9. The design is placed on a human phantom, interaction of radiation waves towards the human body should be considered for analysis. For that specific absorption rate analysis, mainly two standards are considered. The first is the IEEEC95.1-1999 standard [16, 17]. The designed antenna is made of polyimide substrate which is used in medical and body-centric applications. The flexible nature of the substrate with a lower thickness made the ability to curve to the human structures presented in Figure 5.9. The placement of AMC on the back of the antenna has reduced the SAR values to an extent. As per the guidelines, radiation limits should not extend 1.6 W/Kg [18, 19].

There are some fixed standards given by the Federal Communications Commission (FCC), where SAR is about 1.6 W/Kg. The SAR value can be

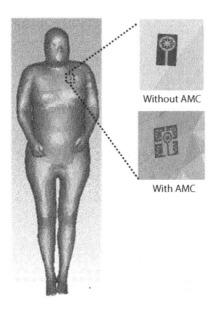

Without AMC

With AMC

Figure 5.9 Star-shaped antenna placed on the human body backed with AMC structure.

calculated by a fundamental equation. Where is 'σ' conductivity and is 'ρ' mass density where E is the electric field [20]

$$SAR = \frac{\sigma |E^2|}{\rho} \qquad (5.2)$$

The SAR analysis on the designed antenna was carried out on CST Microwave Studio software. As the antenna is placed on the chest part radiation effect is observed in Figure 5.10. It represents averaged SAR values over 10gm of human tissue. SAR values are 1.06 W/Kg when the single star-shaped antenna is placed on phantom power of 0.5 W at 5.8 GHz. Similarly, an antenna is backed with AMC structure SAR has lowered to 0.195 W/Kg. By placing antenna backed with AMC structure reduction of SAR values are observed 88% [21, 22].

5.4.2   On-Body Gain of the Star-Shaped Antenna With and Without AMC

The 3D-polar plot distribution on the human body is presented in Figure 5.11. Gain at 5.8 GHz gain is showing the 8.11 dB when the single star-shaped element placed on phantom due to reflection for the body. Similarly,

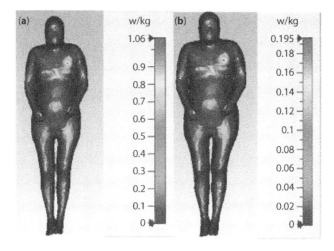

Figure 5.10 Specific absorption rate analysis (a) without AMC (b) with AMC structure.

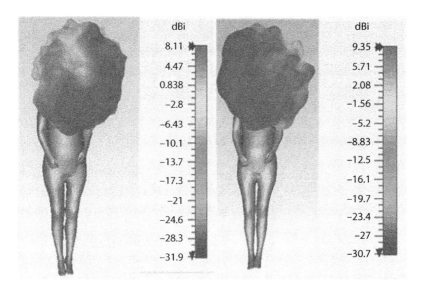

Figure 5.11 On-body gain of the antenna without and with AMC structure.

an antenna is backed with an AMC structure, it provides a gain of 9.35 dB. The gain has improved 1.24 dB by arranging AMC backed structure.

5.4.3   Far-Field Characteristics of An Antenna

The current densities in the entire structure and solves the total far-field solution to the large human body. This structure consists of a huge number

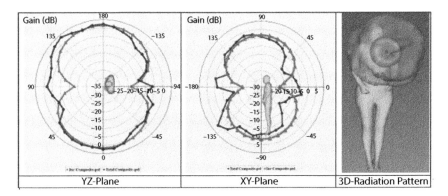

| YZ-Plane | XY-Plane | 3D-Radiation Pattern |

Figure 5.12 Far-field characteristics of antenna in different planes.

of mesh cells with mounted antenna performance. The far-field radiation patterns and human body loaded 3D polar plots are placed in Figure 5.12. The antenna is placed on the human chest which is backed with high dielectric substrates like skin, fat, and muscle, with dielectric constants of 38.01, 5.43, and 55.99, respectively. The antenna placed on the chest shows bidirectional patterns in the YZ-plane. The maximum radiation in the YZ-plane is observed in $180^0/0^0$ at 5.8 GHz. Similarly, at the XY-plane it is observed along with $90^0/-90^0$ at 5.8 GHz frequency.

5.5 Transmitting Signal Strength

To check the transmitting capability of the antenna a small experiment is performed by placing the proposed antenna at the receiving end and transmitting power using AARONIA POWER LOG70180 Horn antenna with

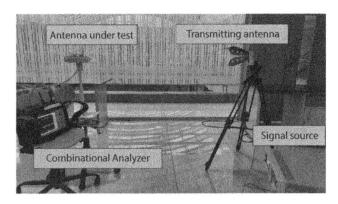

Figure 5.13 Experimental setup for measuring received power of a star-shaped antenna.

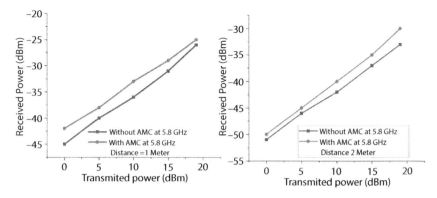

Figure 5.14 Measured performance of received power with and without a human body.

connected Keysight EXG Microwave signal generator shown in Figure 5.13. The Anritsu Combinational analyzer is switched to spectrum mode to estimate the received power. From the horn antenna signal power of 0dBm, 5 dBm, 10 dBm, 15dBm, and 19 dBm is transmitted. At the receiving end received power of the proposed antenna is calculated at different distance points which are 1 m and 2 m. The amplitude of receiving power increases with increasing transmitting power which is shown in Figure 5.14.

5.6   Conclusion

This chapter concentrates on wearable antenna ISM frequencies (5.8 GHz). AMC placed which is having 2 × 2 array elements arranged on the back of the antenna. AMC structure having dimensions of 30 × 30 × 0.2 mm³ and compact. The antenna with AMC structure is placed on the human body structure with defined dielectric constant as of the skin, fat, and muscle structures and the performance is analyzed. By placing AMC, a gain of antenna has improved 1.1 times when compared to monopole structure and SAR value decreased by 88%. The flexible nature of the antenna is analyzed with a shift in the frequency. The antenna has a stable radiation pattern in the far-field environment on the phantom structure.

References

1. Okas, P., Sharma, A., Gangwar, R.K., Super-wideband CPW fed modified square monopole antenna with stabilized radiation characteristics. *Microw Opt Technol Lett.*, 60(3), p.568-575, 2018.

2. Poffelie LAY., Soh, P.J., Yan, S., Vandenbosch, G.A., A high-fidelity all-textile UWB antenna with low back radiation for off-body WBAN applications. *IEEE Trans Antennas Propag.*, 64(2), p.757-760, 2016.

3. Samal, PB., Soh, P.J., Vandenbosch, G., A. UWB all-textile antenna with full ground plane for offbody WBAN communications. *IEEE Trans. Antennas Propag.*, 62(1), p.102-108, 2014.

4. Lee, H., Tak, J., Choi, J., Wearable Antenna Integrated into Military Berets for Indoor/Outdoor Positioning System. IEEE Antennas Wirel. Propag. Lett., 16, p.1919, 2017.

5. Azim, R., Islam, MT., Misran, N., Microstrip line-fed printed planar monopole antenna for UWB applications. *Arabian J Sci Eng.*, 38, p.2415–2422, 2013.

6. Hussin, E. F. N. M., Soh, P. J., Jamlos, M. F., Lago, H., Al-Hadi, A. A., Rahiman, M. H. F., & Lago, M. H. A., Wideband Textile Antenna with Ring Slotted AMC Plane., *Education (MOHE)*, p.527, 2017.

7. Moon, J. I., & Park, S. O., Small chip antenna for 2.4/5.8-GHz dual ISM-band applications. *IEEE Antennas Wirel. Propag. Lett.*, 2, p.313-315, 2013.

8. Bahrami, H., Mirbozorgi, S. A., Ameli, R., Rusch, L. A., & Gosselin, B., Flexible polarization diverse UWB antennas for implantable neural recording systems. *IEEE transactions on biomedical circuits and systems.* 10, p.38-48, 2015.

9. Hazarika, B., Basu, B., & Kumar, J., A multi-layered dual-band on-body conformal integrated antenna for WBAN communication. *AEU-International Journal of Electronics and Communications*, 95, p.226-235, 2018.

10. Hong, Y., Tak, J., & Choi, J., An all-textile SIW cavity-backed circular ring-slot antenna for WBAN applications. *IEEE Antennas Wirel. Propag. Lett.*, 15, p.1995-1999, 2016.

11. Agarwal, K., Guo, Y. X., & Salam, B., Wearable AMC backed near-endfire antenna for on-body communications on latex substrate. *IEEE Transactions on components, packaging and manufacturing technology*, 6(3), p.346-358, 2016.

12. Jiang, Z. H., Brocker, D. E., Sieber, P. E., & Werner, D. H., A compact, low-profile metasurface-enabled antenna for wearable medical body-area network devices. *IEEE Trans. Antennas Propag.*, 62(8), p.4021-4030, 2014.

13. Hazarika, B., Basu, B., & Kumar, J. A multi-layered dual-band on-body conformal integrated antenna for WBAN communication. *AEU-International Journal of Electronics and Communications*, 95, p.226-235, 2018.

14. Prudhvi Nadh, B., Madhav, B. T. P., Siva Kumar, M., Venkateswara Rao, M., & Anilkumar, T., Circular ring structured ultra-wideband antenna for wearable applications. *International Journal of RF and Microwave Computer-Aided Engineering*, 29(4), 2018.

15. Rajagopal, S., Chennakesavan, G., Subburaj, D. R. P., Srinivasan, R., & Varadhan, A., A dual polarized antenna on a novel broadband multi-layer Artificial Magnetic Conductor backed surface for LTE/CDMA/GSM

base station applications. *AEU-International Journal of Electronics and Communications*, 80, p.73-79, 2017.

16. Xu, LJ., Wang, H., Chang Y, Bo Y., A flexible UWB inverted-F antenna for wearable application. *Microw Opt Technol Lett.,* 59(10), p.2514-8, 2017.

17. Gao GP, Hu B, Wang SF, Yang C. Wearable Circular Ring Slot Antenna with EBG Structure for Wireless Body Area Network. *IEEE Antennas Wirel. Propag. Let.*17(3), p.43-47, 2018.

18. Cousin, R., Rütschlin, M., Wittig, T., Bhattacharya, A., Simulation of wearable antennas for body centric wireless communication. *Sens Imaging.*, 16, p.1-6, 2015.

19. Gao, G., Hu B., Wang, S., Yang, C., Wearable planar inverted-F antenna with stable characteristic and low specific absorption rate. *Microw Opt Technol Lett.*, 60, p.876-82, 2018.

20. Tripathi, S., Mohan, A., Yadav, S., A performance study of a fractal UWB antenna for on-body WBAN applications. *Microw Opt Technol Lett.*, 59, 2017.

21. Madhav, BT., Anilkumar, T., Kotamraju, SK., Transparent and conformal wheel-shaped fractal antenna for vehicular communication applications. *AEU-Int J Electron Commun.*, 91, 2018.

22. Priyadarshini, S.J., and Hemanth, D.J., Investigation and reduction methods of specific absorption rate for biomedical applications: A survey. *International Journal of RF and Microwave Computer-Aided Engineering*, 28(3), 2018.

6

Antenna Miniaturization for IoT Applications

Sandip Vijay[1]* and Brajlata Chauhan[2]

*[1]Senior Member IEEE, Director, Tula's Institute, Fellow of ACEEE,
Dehradun, India
[2]DIT University, Dehradun, India*

Abstract

Long-range wireless connectivity is the critical issue for many Internet of Thing (IoT) applications especially for those that need to be mobile. The communication among gadgets and other entities relies on radio wave, that's at risk of many attacks. Size anticipation is one of the questions while thinking about IoT devices, in conjunction with radio performance and charge. IoT is an extremely convoluted heterogeneous organization stage. The antennas used for IoT bundles are needed to demonstrate three essential qualities, as (i) miniature length, (ii) electricity performance and (iii) capability to perform in multi-antenna climate. IoT gadgets show highlights IEEE 802.15.3 A (high-data-rate WPAN) standard such as high data transmission, basic equipment setup, low force utilization, little size, low obstruction, omnidirectional patterns and a direct stage reaction. Latest expertise in 3D-antenna is assumed to overcome some of the disadvantages identified in conventional antennas where it is required for a certain application. Two-dimensional equivalents of volumetric meta-material engineered as Meta-surfaces (MTSs) are to achieve extraordinary electromagnetic properties in 3-dimensions. Dipoles and waveguides horns produced low bandwidth in most of the remote gadgets. Consequently, high data transmission with the smallest antenna measurements with the basic plan is needed for any handheld IoT gadgets.

Keywords: IEEE 802.15.3, meta-surfaces, PIFA antenna, Internet of Things/Everything, framework on-chip

Corresponding author: vijaysandip@gmail.com

Prashant Ranjan, Dharmendra Kumar Jhariya, Manoj Gupta, Krishna Kumar, and Pradeep Kumar (eds.) *Next-Generation Antennas: Advances and Challenges*, (105–118) © 2021 Scrivener Publishing LLC

6.1   Introduction

Miniaturization of the Antenna has been the theme of various investigations for about 80 years. Early investigations demonstrated that reduction in the size of an antenna brings about an immediate decrease in its transmission competence and proficiency (hr). The size limitation changes over into a lower limit on the achievable radiation quality factor (Q factor) and therefore on the most extreme feasible impedance transfer rate. As of now, numerous new examinations have led to a decrease in the structure factor (or the overall size) of different kinds of antennas while endeavoring to maintain satisfactory organizing properties and working transmission capacity. These miniaturization techniques are usually related to changing the electrical and genuine properties of an antenna. There is plenty of distributed research work with antenna miniaturization. Miniaturization techniques and diagram their fundamental ideas and contrasts. Such an investigation would clarify the favorable circumstances and weaknesses of each group and interface each group to true applications [26, 28, 43]. Also, this examination would be valuable for planning conservative and additionally incorporated antenna frameworks for the fields of broadcast communications, medication, industry, public wellbeing, and security. This article gives such an investigation.

   Here, we depict the most unmistakable antenna miniaturization procedures and separate them into two primary classifications: geography and material-based strategies [1]. Every classification comprises a few dissimilar modules, including antennas dependent on wander lines, fractals, space-filling bends, slow-wave structure, designed ground planes, receptive burdens, towering dielectric steady substrates, and meta-materials. It is unimaginable to expect to refer to each distributed work in these classifications. Notwithstanding, we have endeavored to plot each group hidden electromagnetic establishment, favorable circumstances, weaknesses, and pertinent applications through distributed models [18, 19, 36].

   Internet of Things/Everything (IoT/E) will assume a significant part in contemporary correspondence organizations. In this advanced and future correspondence worldview, many diverse shrewd Internet-empowered gadgets will speak with one another consistently. For such correspondence organizations and keen gadgets, aside from the refined correspondence conventions, productive equipment will likewise assume an urgent job. Antennas being at the frontage finish of the correspondence, are one of the significant segments of such equipment. The antennas for

IoT applications are needed to show three significant attributes: (i) small size, (ii) energy production, and (iii) capacity to work in a multi-antenna climate.

The antenna is nowadays used for many medical applications. The antennas utilized within the body need to fulfill numerous circumstances, not at all like conventional antennas which work in free space. Since 1990, microwave sensors have assumed a significant function in IMD; W. Greatbatch and C.F. Homes [2] in 1991 the first inserted electro-biomedical gadget was the heart pacemaker. Presently, a few defibrillators and medication conveyance frameworks are as a rule utilized [31, 35, 39, 40]. Thinking about the need for wider application and public awareness about wellbeing, the IMDs [3, 4] have become a developing innovation. It is utilized for observing the human internal heat level which is presented by W.G.Scanlon, N.E. Evans, and Z.M. Mc Creech, [5] in 1997, inserted defibrillators [6] that reestablish an ordinary heartbeat by sending an electric heartbeat or stun [7] to the heart, D. Wessels [8] in 1992 proposed Implantable pacemakers and defibrillators; T. Yilmaz in [12] 2008 demonstrated constant glucose-level monitoring device; E.Y. Chow [16] in 2009 proposed cardiovascular stunts as an antenna for wireless applications like cardiovascular breakdown location, and so on The point of the IMD, to ceaselessly screen the different exercises within the human body and launch data to a collector exterior to the body, is demonstrated by P. Soontornpipit, C.M. Furse, and Y.C, Chung. Consequently, [20] in 2004 J. Kim, and Y. Rahmat-Samii [27], produced microwave sensors for the utilization of remote correspondence from within the human body to the gadget external. These sensors are insertable antennas; IMDs with a smaller than predictable antenna are known as a radio recurrence (RF)-linked IMD. P. Soontornpipit and C. Gosalia [29] found that such sensors have many difficulties due to the lossy tissues in the human body [32]. J. Kim, C. Gabriel [30] observed that the inserted antenna has certain highlights/qualities, for example, small size, wide working data transmission, and adequate radiation efficiency and needs to address tolerant wellbeing. A few explorations work toward planning an antenna for clinical inserts has been completed in the ISM-band for biometry. As of late, M.I.C.S. has specified 402-405 MHz for this reason to limit obstruction with different administrations. This implies that the frequency is almost 74 cm, which has an immediate connection with the antenna measurements. This shows a need to scale down the microwave antenna which adjusts to the human body. Rules and guidelines identified with all recurrence groups used to plan embedded antennas are given [33, 34, 37, 42].

6.2   Issues in Antenna Miniaturization

One of the basic boundaries to mechanical headways of innovative IoT-related gadgets is rigidity originating from structure factor and weight contemplations. While there have been requests for sizes of advances in miniaturization, adaptability is a component that is difficult to prevail, for various sensors have various sizes and distinctive actual measurements. Hence, there is a need for scaled-down antennas to be incorporated alongside the sensor and without stacking the gadget. Subsequently, application-explicit antennas are intended for singular sensors or gadgets. The antenna plans have difficulties like miniaturization, minimization, and low force. As a more noteworthy number of sensors and circuit gadgets are associated with utilizing IoT it likewise raises power the board for different sensors. In the vast majority of the mechanical and ecological and calculated applications, different sensor hubs are put at distantly enormous distances and controlling them utilizing batteries or different sources is troublesome and unrealistic. A large portion of the sensors is to work on the low force. Low force supply to sensors can be accomplished utilizing RF energy gathering since sensors are encircled by numerous RF sources like GSM/LTE, WI-FI, WIMAX, WLAN, and different sources. In an RF gathering framework, the RF power is extricated utilizing Antennas at that point given to redressing circuits followed by power combiners and conveyed to stack, the amendment is absurd.

Ongoing advancements in designed materials have been utilized to increase the field of adaptable hardware. Adaptable electronic gadgets are frequently lightweight, compact, more affordable, climate agreeable, and expendable. The adaptable gadgets market is required to arrive at 40.37 billion in income by 2023 [21–24]. As we remotely associate an ever-increasing number of gadgets to the Internet, hardware engineers face a few difficulties, including how to bundle a radio transmitter into their current gadget land and how to make progressively more modest gadgets. They're additionally endeavoring to satisfy customer need for Internet of Things (IoT) items that are ergonomically simple to utilize and inconspicuous. Size desire is one of the most posed inquiries while thinking about IoT gadgets, alongside radio execution and cost.

Preferably, architects might want to utilize IoT parts that are as little as could be expected under the circumstances, have incredible RF execution, and are moderate. These qualities don't normally meet in IoT part contributions, and that presents a test for arrangement suppliers. Luckily, the size of a kick bucket has been getting more modest throughout the years as the business receives new fabricating measures. The business has been fathoming the space issue for IoT executions by joining the MCU and RF frontend into framework on-chip (SoC) arrangements (i.e., making remote MCUs accessible.) However, the pattern toward SoCs has not comprehended the material science of the RF transmitter—the antenna. Antenna configuration is frequently left for clients to figure out, or they might be guided to pick prepared to-utilize remote modules with an incorporated antenna. The space needed for an antenna is a test that accompanies planning little IoT gadgets. It should be productive while additionally empowering dependable remote associations. Therefore, the focal point of this section is featuring the particular worries around the antenna combination.

At the point when the IoT began to rapidly develop during the 2000s, the business was called machine-to-machine (M2M), and the parts offered for IoT networks were primarily broad bundle radio assistance (GPRS) modems, Bluetooth sequential link substitution, or sub-GHz exclusive radios. These plans had two principle parts for availability: the MCU and the radio modem. Also, the necessary space for fundamental IoT usefulness was regularly at its littlest—50 mm on each measurement—which means the gadgets were about the size of a cell phone. At the point when the semiconductor business moved to measures where the necessary MCU and RF usefulness could be bundled into a similar pass on, additional opportunities for designers started to arise. Presently designers could execute the usefulness of an IoT gadget in a similar IC/SoC. The IoT part models moved to remote MCUs due to the undeniable advantages—architects could plan IoT gadgets with a solitary segment and spare huge space, yet they could likewise set aside cash given the lower segment costs. While choosing the engineering for current IoT gadgets, it's likely the SoC-based frameworks will lead the way because of their size advantage.

6.3   Antenna for IoT Applications

The new period of profoundly incorporated SoCs leaves engineers with certain inquiries concerning the antenna: In general, how much space would it be advisable to hold for the antenna? Researchers were thinking about what sort of antenna would be a good idea to choose—should we utilize a module with the antenna previously coordinated? The antenna queries are unpredictable at numerous levels, as we need to think about size and proficiency, however detuning inquiries also, particularly across plans that may have various lodgings with a similar antenna engineering. It has been normal to utilize printed circuit board (PCB) follow antennas, for example, altered F, for IoT plans because of their low bill of material (BOM) costs. Yet, these PCB antennas have huge size prerequisites, typically in the scope of 41 mm x 30 mm, eventually making the subsequent IoT gadgets gigantic. These antennas have another disadvantage whenever utilized in a module also: They are touchy to the detuning brought about by the lodging materials and require explicit thought in the item gathering to work ideally. In SoC plans, antenna tuning is important for the ordinary plan stream and requires a specific measure of mastery. In these plans, a printed antenna doesn't contrast from other antenna types except for the size.

Antenna makers have been offering "chip antennas" for a long while to rearrange plan endeavors; however, there are likewise size benefits. These chip antennas are offered in two unique structures; antennas that are not coupled to the ground plane will require a moderately huge leeway territory (or that is liberated starting from the earliest stage, and segments), including monopole and reversed – F type antennas in Figures 6.1 and 6.2. Antennas that are coupled to the ground plane require either a generally little freedom zone underneath the antenna or don't need leeway territory by any means. Both antenna types have space necessities on either freedom zone or potentially ground plane and PCB size. The necessary space for the RF a piece of an IoT configuration ought to likewise incorporate the fundamental leeway territory because no segments or follows can be set here [18, 25].

This implies that when planners are doing measure assessments for IoT gadgets, they need to focus on the vital measurements for the antenna and the necessary leeway zones, yet they likewise need partition good ways from the antenna with the edge of the lodging. IoT will be basic in consolidating distinctive heterogeneous end frameworks including everything from smart homes to the modern Internet of Things with keen agribusiness, shrewd urban areas, and the brilliant lattice. Aside from the complex correspondence conventions, the decision of an antenna framework will be a basic part of all these hub end keen gadgets. Picking the correct antenna for an application forces a key plan challenge. As IoT modules keep on contracting fusing more remote advances, making space for antennas is

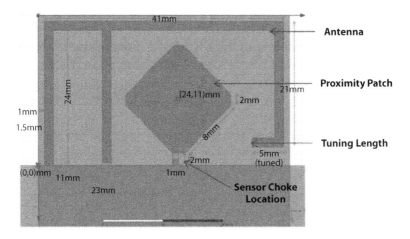

Figure 6.1 Inverted F application.

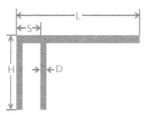

Figure 6.2 Inverted F design layout.

turning into an undeniably huge test. In this manner, IoT-module antenna configuration faces the limitations of steadily contracting impressions while keeping up sensible antenna execution under extreme conditions, for example, clamor, blurring, and the requirement for productivity. Upgraded strategies for multiplexing, impedance alleviation, booking, and radio asset portion work close by the antenna plan for the acknowledgment of effective antenna frameworks for IoT.

The unstable development of the Internet of Things and brilliant modern applications makes numerous logical and designing difficulties that are the principal goals of this unique issue for astute examination endeavors for the improvement of proficient, practical, adaptable, and solid antenna frameworks for IoT. The antenna plans for UWB and RFID labels in IoT empowered climate, MIMO antenna frameworks, transmission strategies of huge MIMO frameworks, and position resilience plan strategy and impacts of irregularity in component positions in antenna arrays.

A lot of studies have been centered around investigating novel added substances and creative innovation for the following innovation remote segments and structures while in transit to be reasonable for IoT gadgets. These gadgets show highlights featured IEEE 802.15.3 A standard, i.e., towering information transmission, straight-forward tools design, low force utilization, small size, low obstruction, Omni-directional radiation drawings, and a direct stage reaction. The information impedance of IC chip differs from one maker to another producer [7, 9–11, 13–15, 17, 18].

6.4   Miniaturize Reconfigurable Antenna for IoT

The first thing which comes to mind about the external antenna of IoT devices: Looking at a few distinctive antenna bundling choices, almost 60% of IoT 2.4 GHz clients assess the exhibition and plausibility of the outside antenna (antennas coordinated into the lodging model through U.Fl connector). In any case, roughly 8-10% of these assessed plans send the outer antenna, and 92-90% of the clients pick modules with an underlying chip antenna. For what reason would engineers not broadly convey outer antennas on their plans? The appropriate response has two fundamental measurements. To start with, the mechanics of an outside antenna does not plan inviting; they look terrible and break effectively if the IoT gadget is dropped. These outside antennas additionally altogether increment BOM and get together expenses. Additionally, when contrasting the proficiency of an all-around constructed RF plan with a chip antenna versus an outer antenna through a U.Fl antenna connector, there is no advantage in utilizing an outside antenna. The advantage of the outer antenna is self-evident, if the lodging of the gadget is metallic, framing a faraday confine that makes it unthinkable for the RF signs to infiltrate the gadget. An outer antenna is additionally defended if the most flawlessly awesome execution is required and get together expenses and mechanical plans consider its utilization.

When designing the IoT gadget with an antenna, the mechanics and lodging assume a critical function in keeping away from or causing antenna detuning. The RF radiation, when blasting out of the antenna, is affected by the closeness of the materials. The antenna will detune on the off chance that it contacts the metal or even plastic. Hence, the antenna should be isolated from actual contact with lodging plastic or metal. There are enormous contrasts in the kinds of antennas and their susceptibility to detuning. Monopole-type antennas are more touchy than ground-coupled antennas. It unreservedly on their plans, lessening the size of the gadgets essentially. When planning little IoT gadgets the size of a coin cell battery, there is consistently a trade-off with antenna effectiveness. The more modest the size we attempt to accomplish, the less proficiency we can have for the RF execution. Gadgets under 10 mm on each measurement start to accomplish execution in the 2.4 GHz band, giving clients Bluetooth availability of roughly 10 meters with a cell phone, which is adequate for most close to home IoT gadgets. Notwithstanding, when the measurements are more like 20 mm toward every path, the RF productivity increments altogether, giving a useful scope of 20–40 meters with a cell phone contingent upon the conditions.

In the integration of the antenna (IoA) module, it doesn't de-tune, even with the vicinity of the final result mechanics. This powerful antenna engineering permits the arrangement of the module close by the item lodging, wiping out the requirement for the complete 3-dimensional leeway zone with the finished result lodging. This element permits the planner to put the joining module on minimized plans without investing loads of energy in antenna detuning and improvement. Now the question arises, why Module vs. SoC [41]? In RF modules, the difficult undertaking of antenna coordinating has been finished with no application-explicit tuning essential by and large. Notwithstanding, now and again the item configuration can't keep away from the position of a connection screw, LCD, or battery near the antenna, causing huge changes in the antenna full recurrence and general execution. The modules themselves can't be handily adjusted to make up for the detuning. When utilizing an SoC with an antenna coordinated for the real end-application plan, an accomplished architect can meet the necessary execution and cost objectives. The streamlining of antenna execution and the SoC RF format requires a comprehension of RF configuration rules and very costly gear. As the quantity of remote applications and gadgets builds, the necessary administrative confirmation may turn out to be excessively costly and tedious. This is particularly evident when the item portfolio increases to several or even tens of gadgets, each requiring separate affirmation related item the executives. An undeniably mainstream arrangement is to utilize pre-affirmed, prepared-to-utilize remote modules, particularly for IoT-related plans.

6.5   Conclusion & Future Work

The most recent ability in 3D-antenna is accepted to beat a portion of the disservices distinguished in traditional antennas where it is needed for certain applications. Two-dimensional reciprocals of volumetric metamaterial designed as Metasurfaces (MTSs) are to accomplish remarkable electromagnetic properties in 3-measurements. Various most recent readings have gotten the reconfigurability together with metasurfaces to acknowledge multi-dimensional advantages trying to accomplish strong fields, more modest size, and expanded controllability. In the relation to planning trademark, an antenna as a front part of any remote gadget is needed to have a wide transfer speed, great radiation design, direct stage reaction, and switchable capacity. Prior, tight data transmission highlights of microstrip fix antenna were a genuine undertaking for specialists. It was depicted that previous restricted data transmission highlights were a significant constraint of 13 to 52% of ordinarily utilized antennas, for example, dipoles and waveguides horns in the greater part of the remote gadgets. There are numerous difficulties for scientists to represent considerable authority in while planning antenna. In current correspondence, most extreme antennas are multifaceted in plan with lopsided patches and dielectric substrates. The highlights of such kinds of antennas can likewise be influenced by the math of heightening gadgets on which antennas are being found. Henceforth, high transmission capacity with the littlest antenna measurements with a straightforward plan is needed for any handheld IoT gadgets.

References

1. Wheeler H. A., "Fundamental limitations of small antennas," in *Proc. IRE*, vol. 35, no. 12, pp. 1479–1484, Dec. 1947.
2. Greatbatch, W. and Homes, C.F., "History of implantable devices", *IEEE Engineering in Medicine and Biology Magazine: The Quarterly Magazine of the Engineering in Medicine & Biology Society*, Vol. 10 No. 3, pp. 38-41, 1991.
3. R. C. Hansen, "Fundamental limitations in antennas," *Proc. IEEE*, vol. 69, pp. 170–182, Feb. 1981.
4. J. S. McLean, "A re-examination of the fundamental limits on the radiation Q of electrically small antennas," *IEEE Trans. Antennas Propagation.*, vol. 44, pp. 672–676, May 1996.
5. Scanlon, W.G., Evans, N.E. and Mc Creesh, Z.M., "RF performance of a 418MHz radio telemeter packed for human vaginal placement", *IEEE Transactions on Biomedical Engineering*, Vol. 44, No. 5, pp. 427-430, 1997.

6. J. Rashed and C. T. Tai, "A new class of resonant antennas," *IEEE Trans. Antennas Propag.*, vol. 39, no. 9, pp. 1428–1430, Sept. 1991.

7. C. R. Medeiros, J. R. Costa, and C. A. Fernandes, "RFID reader antennas for tag detection in self-confined volumes at UHF," *IEEE Antennas Propag. Mag.*, vol. 53, no. 2, pp. 39–50, Apr. 2011.

8. Wessels, D., "Implantable pacemakers and defibrillators: device overview and EMI considerations", *Proceedings of the IEEE International Symposium on Electromagnetic Compatibility (EMC2002)*.

9. K. V. S. Rao, P. V. Nikitin, and S. F. Lam, "Antenna design for UHF RFID tags: A review and a practical application," *IEEE Trans. Antennas Propag.*, vol. 53, no. 12, pp. 3870–3876, Dec. 2005.

10. Alien Technology LLC. Alien. [Online]. Available: http://www.alientechnology.com

11. H. Makira, Y. Watanabe, K. Watanabe, and H. Igarashi, "Evolutional design of small antennas for passive UHF-band RFID," *IEEE Trans. Magn.*, vol. 47, no. 5, pp. 1510–1513, May 2011.

12. Yilmaz, T., Karacolak, T. and Topsakal, E., "Characterization and testing of a skin mimicking material for implantable antennas operating at ISM band (2.4 GHz–2.48 GHz)", *IEEE Antennas and Wireless Propagation Letters*, Vol. 7, pp. 418-420, 2008.

13. D. H. Werner and S. Ganguly, "An overview of fractal antenna engineering research," *IEEE Antennas Propag. Mag.*, vol. 45, no. 1, pp. 38–57, Feb. 2003.

14. S. R. Best, "On the resonant properties of the Koch fractal and other wire monopole antennas," *IEEE Antennas Wireless Propag. Lett.*, vol. 1, pp. 74–76, 2002.

15. K. J. Vinoy, J. K. Abraham, and V. K. Varadan, "On the relationship between fractal dimension and the performance of multi-resonant dipole antennas using Koch curves," *IEEE Trans. Antennas Propag.*, vol. 51, no. 9, pp. 2296–2303, Sept. 2003.

16. Chow, E.Y., Ouyang, Y., Beier, B., Chappell, W.J. and Irazoqui, P.P., "Evaluation of cardiovascular stents as antennas for implantable wireless applications", *IEEE Transactions on Microwave Theory and Techniques*, Vol. 57, No. 10, pp. 2523-2532, 2009.

17. P. Lande, D. Davis, N. Mascarenhas, F. Fernandes, and A. Kotrashetti, "Design and development of printed Sierpinski Carpet, Sierpinski Gasket and Koch Snowflake fractal antennas for GSM and WLAN applications," in *Proc. Int. Conf. Technology Sustainable Development (ICTSD)*, pp. 1–5, 2015.

18. N. A. Saidatul, A. A. A.-H. Azremi, R. B. Ahmad, P. J. Soh, and F. Malek, "Multiband fractal planar inverted F antenna (f-PIFA) for mobile phone application," *Prog. Electromagn. Res. B*, vol. 14, pp. 127–148, 2009.

19. J. M. Gonzalez-Arbesu, S. Blanch, and J. Romeu, "Are space-filling curves efficient small antennas?" *IEEE Antennas Wireless Propag. Lett.*, vol. 2, no. 1, pp. 147–150, 2003.

20. Soontornpipit, P., Furse, C.M. and Chung, Y.C., "Design of implantable microstrip antennas for communication with medical implants", *IEEE Transactions on Microwave Theory and Techniques*, Vol. 52, No. 8, pp. 1944-1951, 2004.

21. S. K. Mahto, A. Choubey, and R. Kumar, "A novel compact multi-band double Y-slot microstrip antenna using EBG structure," in *Proc. Int. Conf. Microwave Photonics (ICMAP)*, pp. 1–2, 2015.

22. D. Guha, M. Biswas, and Y. M. Antar, "Microstrip patch antenna with the defected ground structure for cross-polarization suppression," *IEEE Antennas Wireless Propag. Lett.*, vol. 4, no. 1, pp. 455–458, 2005.

23. A. K. Arya, M.V. Kartikeyan, and A. Patnaik, "On the size reduction of microstrip antenna with DGS," in *Proc. 35th Int. Conf. Infrared Millimeter Terahertz Waves (IRMMW-THz)*, pp. 1–3, Sept. 2010.

24. P. L. Chi, R. Waterhouse, and T. Itoh, "Antenna miniaturization using slow-wave enhancement factor from loaded transmission line models," *IEEE Trans. Antennas Propag.*, vol. 59, no. 1, pp. 48–57, Jan. 2011.

25. C. R. Rowell and R. D. Murch, "A capacitively loaded PIFA for compact mobile telephone handsets," *IEEE Trans. Antennas Propag.*, vol. 45, no. 5, pp. 837–842, May 1997.

26. R. Azadegan and K. Sarabandi, "A novel approach for miniaturization of slot antennas," *IEEE Trans. Antennas Propagation.*, vol. 51, no. 3, pp. 421–429, Mar. 2003.

27. Kim, J. and Rahmat-Samii, Y., "Implanted antennas inside a human body: simulations, designs, and characterizations", *IEEE Transactions on Microwave Theory and Techniques*, Vol. 52 No. 8, pp. 1934-1943, 2004.

28. N. Behdad and K. Sarabandi, "Slot antenna miniaturization using distributed inductive loading," in *Proc. IEEE Antennas Propagation Society Int. Symp.*, vol. 1, pp. 308–311.

29. Soontornpipit, P., Furse, C.M. and Chung, Y.C. (2005), "Miniaturized biocompatible microstrip antenna using genetic algorithm", *IEEE Transaction Antennas Propagation*, Vol. 53 No. 6, pp. 1939-1945, 2003.

30. Kim, D.-H., Kim, D. and Viventi, H.J., "Dissolvable films of silk fibroin for ultrathin conformal bio-integrated electronics", *Nature Materials*, Vol. 9, No. 6, pp. 511-517, 2010.

31. Kaur, G., Kaur, A., Toor, G., Dhaliwal, B. and Pattnaik, S., "Antennas for biomedical applications", *Biomedical Engineering Letters*, Vol. 5, No. 3, pp. 203-212, 2015.

32. Gabriel, C., Gabriel, S. and Corthout, E., "The dielectric properties of biological tissues: an i. Literature survey", *Physics in Medicine and Biology*, Vol. 41, pp. 2231-2249, 1996.

33. Emami-Nejad, H. and Mir, A., "Design and simulation of a flexible and ultra-sensitive biosensor based on frequency selective surface in the microwave range", *Opt Quantum Electron*, Vol. 49, No. 10, p. 320, 2017.

34. Asili, M. ,"Flexible microwave antenna applicator for Chemo – Thermotherapy of the breast", *IEEE Antennas Wireless Propagation Letters*, Vol. 14, pp. 1778-1781, 2015.

35. Aleef, T.A., Hagos, Y.B., Minh, V.H., Khawaldeh, S. and Pervaiz, U., "Design and simulation-based performance evaluation of a miniaturized implantable antenna for biomedical applications", *Micro & Nano Letters*, Vol. 12, No. 10, pp. 821-826, 2017.

36. H Herth, E., Guerchouche, K., Rousseau, L., Calvet, L.E. and Loyez, C., A Biocompatible and Flexible Polyimide for Wireless Sensors, Etienne Springer-Verlag Berlin Heidelberg, *Microsyst Technol*, pp. 1-9, 2017.

37. Huang, Y., Alrawashdeh, R. and Cao, P. , "Flexible meandered loop antenna for implants in MedRadio and ISM bands", *Electronics Letters*, Vol. 49, No. 24, pp. 1515-1517, 2013.

38. Yilmaz, T., Karacolak, T. and Topsakal, E., "Characterization and testing of a skin mimicking material for implantable antennas operating at ISM band (2.4 GHz– 2.48 GHz)", *IEEE Antennas and Wireless Propagation Letters*, Vol. 7, pp. 418-420, 2008.

39. Scarpello, M.L., "Design of an implantable slot dipole conformal flexible antenna for biomedical applications", *IEEE Transactions on Antennas and Propagation*, Vol. 59, No. 10, pp. 3556-3564, 2011.

40. Bahrami, H., Porter, E. and Santorelli, A., "Flexible sixteen antenna array for microwave breast cancer detection", *IEEE Transactions on Biomedical Engineering*, Vol. 62, No. 10, pp. 2516-2525, 2015.

41. Yang, S., Liu, P., Yang, M., Wang, Q., Song, J. and Dong, L. , "From flexible and stretchable Meta-Atom to metamaterial: A wearable microwave meta-skin with tunable frequency selective and cloaking effects", *Scientific Report*, Article number:21921, Vol. 6, 2016.

42. CEPT/ERC 70-03, "European radio communications commission (ERC) recommendation 70-03 relating to the use of short-range", *European Conference of Postal and Telecommunications Administration*, CEPT/ERC 70-03, Annex 12, 1997.

43. R. O. Ouedraogo, E. J. Rothwell, A. R. Diaz, K. Fuchi, and A. Temme, "Miniaturization of patch antennas using a metamaterial-inspired technique," *IEEE Trans. Antennas Propagation*, vol. 60, no. 5, pp. 2175–2182, May 2012.

7

Modified Circular-Shaped Wideband Microstrip Patch Antenna for Wireless Communication Utilities

Manpreet Kaur[1]*, Jagtar Singh Sivia[1] and Navneet Kaur[2]

[1]YCoE, Punjabi University Guru Kashi Campus,
Talwandi Sabo, Punjab, India
[2]Punjabi University, Patiala, Punjab, India

Abstract

This chapter illustrates the design and characterization of a wideband circular patch antenna for wireless communication utilities. It begins with a brief introduction to wireless communication and microstrip antenna, followed by a comprehensive literature review of few antennas designed for same type of applications. A commercially available antenna design tool, named as high-frequency structure simulator (HFSS) is utilized in this work. The designed circular patch of radius 12 mm is placed on the upper surface, whereas a rectangular ground plane of size 29 x 32 mm2 is placed at the lower surface. In the realized structure, two similar circular patterns are introduced onto the radiating element along with a partial ground plane. Initially, an antenna with an upper pattern (Antenna 1) is designed and its performance is scrutinized. After that, an antenna with a lower pattern (Antenna 2) is analyzed, At the end, both the patterns are combined in one structure (Projected Antenna). From the results, it is depicted that resonances appear at 5.52, 11.36, 15.76, and 18.05 GHz with an associated S11 value -17.78, -29.04, -20.96, and -21.21 dB. Moreover, the influence of variations in antenna design specifications and different substrate materials is also demonstrated for deep understanding.

Keywords: Circular patch antenna, multiband behavior, S11 characteristics, similar circular patterns, gain, radiation pattern, HFSS, current distribution

**Corresponding author*: sketty@rediffmail.com

Prashant Ranjan, Dharmendra Kumar Jhariya, Manoj Gupta, Krishna Kumar, and Pradeep Kumar (eds.)
Next-Generation Antennas: Advances and Challenges, (119–142) © 2021 Scrivener Publishing LLC

7.1   Overview of Wireless Communication

In this modern era, the wireless communication sector includes different types of specifically defined methods and techniques for providing reliable communication between the devices [1]. The term wireless communication was first introduced in the 19th century and has been in the developing phase ever since. It also incorporates a wide variety of services and applications to satisfy user-related requirements [2]. The evolution of this technology has brought many improvements with notable characteristics. It enables people to communicate regardless of their location. It also allows easy accessibility as the remote areas are fully connected [3]. Wireless communication uses infra-red, radio-frequency, acoustic, and any other type of electromagnetic waves in place of cables or wires to transmit data. This is in contrast to a wired network, which is a bounded network that allows communication over a limited range [4]. The setup and infrastructure required for wireless systems are simple and easier than wired communication systems. Therefore, these evolving wireless technologies are becoming more popular these days. In this sector, efficient antennas are employed as they are potentially responsible for the enhancement in system performance [5]. Nowadays, modern wireless communication technology, being the core technology, requires miniaturized, efficient, and high-performance antennas that allow communication over remotely operated areas [6]. These antennas have to perform critical activities that are required for efficient communication. In the past decade, different strategies have been proposed by distinct researchers in the antenna community. Design specifications of the structures are the main factors for the computation of fundamental frequencies [7]. The demand for high-fidelity antennas with wider impedance bandwidth and rotatable radiation patterns is also increasing day by day. The antennas with high gain values can improve the signal strength to a large extent [8]. Therefore, they have been successfully utilized in various sectors that include cellular communication systems, satellite communication links, military, defense, and health care services, and so on [5]. The design of a high-capacity, multi-functional antenna based on ambitious requirements is a challenging task for antenna designers.

7.2   Introduction to Microstrip Patch Antenna

The purpose of an antenna is to send or receive electromagnetic waves. It is an arrangement of single or multiple conductors and is commonly used in an open space. Several types of antennas are available but in this proposed work, a microstrip antenna is discussed. Due to low and planer profiles, microstrip antennas have gained significant research attention in the last few decades. These antennas can be incorporated onto the surface of aircraft and missiles. In the antenna design theory, the most forward-looking topic for research is the microstrip patch antenna. The idea of a microstrip antenna was first recommended by G. A. Deschamps in 1953 [9]. After that, Bob Munsan proposed a design in 1972 that was realized practically. The microstrip antennas are formed by adopting different geometrical shapes and dimensions but four of them are the most common shapes – square, circular, rectangular, and hexagonal [10]. These antennas have the capability of transmission and reception over long distances. Therefore, they are employed in a broad range of modern applications due to their easiness in the designing process and compatibility with printed-circuit technology [11]. In general, the structure of a microstrip antenna comprises a thin sheet, which is made up of insulating material. The ground is printed on one side, whereas the design of the patch is patterned on the other side. The edges of the patch experience fringing effects due to finite dimensions along both sides that allow the antenna to radiate [9]. This effect depends on the patch dimensions and thickness of the substrate. Due to this, the size of the patch appears to be more than that of its actual dimensions. The dimensions of the patch are represented as the width (W) and length (L). The substrate has a thickness (h) and a permittivity (ε_r). The length (L) of the patch follows this necessary condition, i.e., $\lambda o /3 < L < \lambda o /2$ and the associated height of the patch is usually $<<\lambda o$ [10]. The most commonly used conducting materials for the patch is gold or copper. The feed lines and the radiating patch are usually embedded on the dielectric substrate by using a photo etching process [12]. A higher dielectric constant (<12) must be used for implementing a compact size antenna but it results in lower efficiency and narrower bandwidth. But, there exists some type of compromise between the antenna dimensions and the performance. The generated electric fields allow the production of radiations from the patch antenna. The radiated electromagnetic waves are of three types. In the initial part, "useful" radiations are radiated into space [9]. The reflected waves that arrive at the region between the patch and ground plane are termed as diffracted waves, which are also responsible for actual power transmission. The terminating part of the wave is trapped in the substrate which is due to total reflection at the air-dielectric separation surface [11].

7.3 Literature Review

In previous years, huge efforts have been made to investigate the analytical behavior of these microstrip antennas. Murugan *et al.* [13] had illustrated the functioning of a square patch-based circularly polarized antenna. The truncated structure was excited with coaxial feed and provided an impedance bandwidth of 32% at the claimed frequency. A microstrip-line-fed structure was proposed in which a circular-shaped defect was introduced in the ground. Based on the mathematical equations, theoretical analysis was done and suggested its usage for the Ku band. The observed gain was in the range of 5 dBi to 12.08 dBi [14]. Deshmukh *et al.* [15] had explained the operation of E-shaped antennas that yielded again more than 3 dBi at the examined resonances. Antennas were dual-polarized and constructed on glass epoxy material. A dual slant polarized geometry printed on two sides was presented and excited with two microstrip feed lines. Investigations revealed that V-shaped antennas exhibited a wideband of much larger bandwidth [16]. Nasimuddin *et al.* [17] had suggested an antenna that was circularly polarized and had an asymmetrically slotted patch. The embedded slots were of different shapes and their results were compared. A rectangular microstrip antenna was implemented with PTFE substrate and a defected ground plane in which four wide slots were etched. The performance was justified by conducting a theoretical analysis [18]. Shanmuganantham *et al.* [19] had designed a probe-fed compact antenna in which two slots were introduced. One is inverted U-shaped and the other is W-shaped. The provided results showed significant size reduction and bandwidth enhancement. A novel and the high gain hexagonal-shaped antenna was proposed in which multiple resonances were practically observed. The compactness computed was 57% and also proved its applicability for wireless applications [20]. Karia *et al.* [21] had discussed a wideband antenna of size 50 x 60 x 1 mm^3. Results were validated by experimentally testing the antenna. Essential antenna parameters were described and current distribution plots were also presented. Computations were done in the range of 30 kHz–14 GHz. A square fractal antenna based on gap coupling was described for bandwidth enhancement. At the operational frequency, the antenna had offered an impedance bandwidth of 85.42%. The modified fractal structure is applicable for Bluetooth, WLAN, and WiMAX applications [22]. A monopole antenna in the form of a rectangular ring was illustrated for WLAN applications. The antenna of size 21.4 x 59.4 mm^2 provided a dual-band operation.

The first band was from 2.21-2.70 GHz and the second band was from 5.04-6.03 GHz. Additionally, the radiation patterns were nearly omnidirectional [23]. Vivel *et al.* [24] had demonstrated the design procedure of FR4 based compact antenna for 2.5/5.7 GHz applications. The area of the antenna was 18.2 x 20 mm^2. Experimental analysis was performed to validate the simulated values. A miniaturized prototype of size 20 x 20 mm^2 was physically realized for wireless services. The geometry was dual-polarized and exhibited significant electromagnetic properties [25]. A modified circular monopole antenna was mentioned for various wireless services. The circular patch had an optimal value of 25 mm. Two circles of radius 12.5 mm were etched from the opposite sides. By the insertion of a triangular notch in the ground, bandwidth enhancement was also examined [26]. Kumar *et al.* [27] had evaluated the behavior of CPW fed antennas. The selected dimensions of these antennas were 70 x 70 mm^2. Resonances were evaluated at 2.45 and 5.51 GHz, with S11 ≤ -10 dB. The performance was analyzed with HFSS software. The structure was composed of a rectangular radiating element in which an arc-shaped strip was etched, a ground plane, and an FR4 substrate. The embedded arc-shaped strip had helped to control the impedance bandwidth. An almost similar gain was observed in the entire range, except at around 6 GHz [28]. Jung *et al.* [29] had constructed a broadband meander-line structure with an impedance bandwidth of 785 MHz. The compact antenna of size 12 x 23.1 mm^2 was manufactured with F-PCB (polyimide) material and was excited with a coaxial probe. The ground plane had a size of 45 x 50 mm^2. Kumar *et al.* [30] had produced a UWB system in which two identical triangular-shaped conducting elements were used. Tapered feed was used for excitation. The overall structure of size 46 x 32.6 x 1.6 mm^3 had demonstrated its usage for MIMO/ diversity applications. The evaluated diversity gain was near 10 dBi. A reduced-size circular patch antenna loaded with metamaterial is described. Before loading, resonance was observed at 6.11 GHz, whereas after loading, the antenna resonated at nearly 6.11 GHz with reduced size. The reduction achieved was almost 64% and the gain achieved was 5.04 dB [31]. Wang *et al.* [32] had proposed new techniques for the enhancement of the front-to-back ratio of the designed antenna. By the application of stubs in the structure, the additional capacitance was introduced. By using the optimized structure, wideband unidirectional radiation characteristics were examined along with reduced size. So *et al.* [33] had utilized parasitic shorting pins for the implementation of the patch antenna. The operating properties were controlled with the employed parasitic strips. Results showed that the designed antenna had sufficiently reduced the size and back radiations. The performance was examined with a meandering shorting strip.

The fundamental rule for designing an antenna for wireless purposes is the selection of proper radiator shape and associated design specifications. The design and application of broadband, multifunctional, and compact antenna have gained overwhelming interest from several researchers in recent years [21]. In this view, the work presented in this chapter is to implement and analyze a compact, modified circular patch antenna for wireless services with some added features. This performed work provides concise and detailed information to the users. The chapter is mainly organized into seven sections. After presenting a brief overview of wireless communication and microstrip antennas in Sections 7.1 and 7.2, a comprehensive literature review on several antennas implemented for similar applications is demonstrated in Section 7.3. The detailed design procedure of the Projected Antenna is illustrated in Section 7.4. The Projected Antenna performance's analysis covering the most relevant information is depicted in Section 7.5. A parametric study is mentioned in Section 7.6 that illustrates how the design parameters affect the device performance. Finally, the complete work is outlined in Section 7.7, mentioning that the Projected Antenna is the best choice for the above-said applications.

7.4 Design and Implementation of Projected Antenna

Circular-shaped patch antennas have been widely employed and exhibit vast and diverse applications. Several teams of researchers have been doing work on several shapes. Here, a circular shape is adopted, as this is the second most popular shape. In this section, a modified circular-shaped antenna is implemented that exhibits multiband operation. Focussing on the prime aim of achieving wideband functionality along with wireless utilities, few alterations are carried out onto the circular patch.

The basic shape considered for the conducting element is a circle. The selected shape is further modified by embedding two symmetrical patterns in the vertical direction. The radius of the basic circle employed in the design is formulated from the basic formula mentioned in Eqns. (7.1, 7.2). The evolution stages of the Projected Antenna is revealed in Figure 7.1. In Figure 7.1(a), a circular pattern is introduced at the upper part of the radiator and is referred to as Antenna-1. In Figure 7.1(b), a circular pattern is positioned at the lower part, which is termed as Antenna-2. Both the patterns are identical in shape and size. In the final shape as shown in Figure 7.1(c), both the patterns are combined in one structure. The designed radiator is further attached with the microstrip feed of dimensions 'F$_L$'=6mm and 'F$_W$' = 3 mm. The feed is located on the left side of the substrate. At this position, suitable impedance matching is observed. The whole arrangement is placed at the upper surface of square-shaped FR4 epoxy material of area 32 x 32 mm^2. The optimal thickness of the FR4 substrate is 1.6 mm [35]. In all the antennas (Antenna 1, Antenna 2, Projected Antenna), ground plane dimensions are the same, i.e. 'L$_{PG}$' = 29 mm and 'W$_{PG}$' = 32 mm. The patterns on the patch are symmetrically positioned, thereby making the overall structure appealing as well as attractive. As a result, multi-resonance behavior is examined with the use of an abovesaid designed patch on one side and the partial ground plane on the other side. The schematic configuration of the Projected Antenna is revealed in Figure 7.2.

The radius of the circular patch is computed by the mathematical formula, which is mentioned below [9]:

$$a = \frac{F}{\left\{1 + \dfrac{2h}{\pi \varepsilon_r F}\left[ln\left(\dfrac{\pi F}{2h}\right) + 1.7726\right]\right\}^{1/2}} \tag{7.1}$$

$$\text{Where } F = \frac{8.791 \times 10^9}{f_r \sqrt{\varepsilon_r}} \tag{7.2}$$

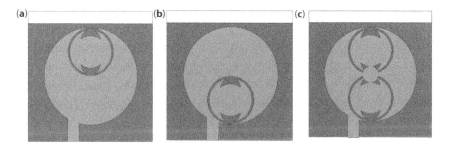

Figure 7.1 Evolution stages (a) Antenna-1 (b) Antenna-2 (c) Projected Antenna.

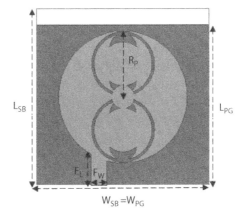

Figure 7.2 Schematic of Projected Antenna.

In these equations, f_r = working frequency, ε_r = relative permittivity, h = selected FR4 substrate's height. Assumptions are: 'f_r' and 'h' are in hertz (Hz) and millimeter(mm), respectively.

7.5   Results and Discussion

For easy and better understanding, the proposed design concept is realized and its operational performance is investigated. Several antenna parameters are examined and are briefly discussed in this section. Simulated results of these parameters are obtained with an electromagnetic 3-Dimensional simulator, Ansoft high-frequency structure simulator (HFSS) software within the range of 3-21 GHz. Figure 7.3 shows the antenna design in the adopted software.

7.5.1   Scattering Parameters (S_{11})

This is one of the most usable parameters regarding antenna performance evaluation. These parameters illustrate that how the input-output terminals associated with an electrical framework or the ports associated with the transmission line are related to each other. Therefore, they represent reflection/transmission characteristics in the frequency domain [9, 34]. S11 parameter is also named as reflection coefficient. Antenna 1 resonates at seven different frequency bands. The resonating frequencies are 6.13, 7.75, 9.92, 12.45, 14.01, 15.16, and 18.35 GHz. The associated S11 value

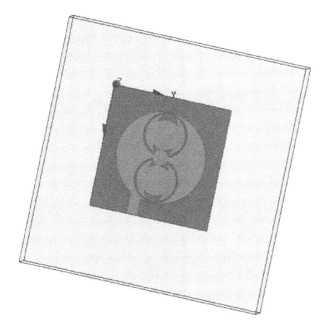

Figure 7.3 View of Projected Antenna in HFSS software.

is -27.83, 11.87, -11.75, -15.59, -23.95, -21.98, and 25.90 dB. In Antenna 2, there are four functional bands. The claimed frequencies are 4.92, 8.89, 16.12, and 17.20 GHz with corresponding S11 value -17.03, -13.36, -19.31, and 21.38 dB. The Projected Antenna offers four bands with center frequencies 5.52, 11.36, 15.76, and 18.05 GHz. The S_{11} values at these frequencies are -17.78, -29.04, -20.96, and -21.21 dB, respectively. Figure 7.4 delineates the S_{11} characteristics of Antenna 1, Antenna 2, and Projected Antenna. Table 7.1 reports the simulated results of the antenna at different stages (Antenna 1, Antenna 2, Projected Antenna).

7.5.2   Voltage Standing Wave Ratio

Voltage Standing Wave Ratio (VSWR) is the key indicator of how efficiently the RF power is delivered from the power source to the antenna with the help of a transmission line [34]. It is a dimensionless quantity. A value of less than 2 is considered a suitable value for several antenna applications [35]. At all the operating frequencies of Antenna 1, Antenna 2, and Projected Antenna, the value of VSWR is in the desirable range.

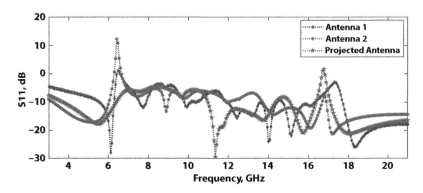

Figure 7.4 S_{11} characteristics of Antenna 1, Antenna 2, and Projected Antenna.

Antenna 1 shows the VSWR value of 1.24, 1.87, 1.01, 1.65, 1.37, 1.69, and 1.22 at 6.13, 7.75, 9.92, 12.45, 14.01, 15.16, and 18.35 GHz, respectively. Antenna-2 yields VSWR values of 1.08, 1.83, 1.45, and 1.76 at the respective frequencies, whereas the Projected Antenna provides VSWR < 2 at the implemented frequencies. All the values are listed in Table 7.1.

7.5.3   Bandwidth

The term bandwidth identifies the frequency range over which an antenna meets a certain set of specifications. Another simple way to define bandwidth is a "range of frequencies around the fundamental frequency where the desired antenna performance characteristics are within an acceptable range" [36]. From Table 7.1, it is noted that Antenna 1 provides operational bandwidth of 12%, 4.9%, 4.2%, 13%, 5.9%, 10.4% and > 16.8% at the respective realized frequencies. Antenna 2 shows a bandwidth range of 3.12-6.07, 8.77-9.02, 13.71-16.72, and 16.90 to more than 21 GHz at the respective functional bands. It is worth mentioning that in the Projected Antenna, the resultant operational bandwidth is 54%, 26%, 8%, and more than 20.5% at the investigated frequencies. It is found that Projected Antenna yields higher bandwidth than other designed antennas (Antenna 1 and Antenna 2).

7.5.4   Gain

Antenna gain is another vital measure that illustrates the operational performance in a better way. This metric represents the capability or level to

Table 7.1 Simulated results of the antenna at different stages (Antenna 1, Antenna 2, Projected Antenna).

Antenna type	'F$_L$', GHz	'F$_C$', GHz	'F$_H$', GHz	Operational bandwidth, %	S11, dB	Gain, dB	VSWR
Antenna 1	5.50	6.13	6.25	12	-27.83	13.57	1.24
	7.52	7.75	7.90	4.9	-11.87	2.85	1.87
	9.80	9.92	10.22	4.2	-11.75	-1.23	1.01
	11.91	12.45	13.59	13	-15.59	6.49	1.65
	13.59	14.01	14.43	5.9	-23.95	4.42	1.37
	14.74	15.16	16.36	10.4	-21.98	6.58	1.69
	17.74	18.35	-	> 16.8	-25.90	2.66	1.22
Antenna 2	3.12	4.92	6.07	64	-17.03	2.54	1.08
	8.77	8.89	9.02	2	-13.36	1.77	1.83
	13.71	16.12	16.72	19	-19.31	7.83	1.45
	16.90	17.20	-	> 21	-21.38	8.25	1.76
Projected Antenna	3.66	5.52	6.43	54	-17.78	11.23	1.16
	11.00	11.36	14.37	26	-29.04	5.55	1.28
	15.04	15.76	16.42	8	-20.96	6.17	1.73
	17.08	18.05	-	> 20.5	-21.21	6.49	1.20

which an antenna concentrates the radiated energy or power in a particular direction, or captures the incident power from that particular direction, in comparison to the lossless isotropic radiator [9, 37]. The minimum required value of antenna gain is 3 dB. The resultant 3-Dimensional plots of Projected Antenna at the claimed frequencies are presented in Figure 7.5. Figure 7.6 shows the comparison of examined gains of Antenna 1, Antenna 2, and Projected Antenna.

7.5.5   Radiation Pattern

The radiation pattern is the pictorial representation of the distribution of radiated energy in the space, relative to the directional coordinates [37]. In other words, it is a vital antenna parameter that tells us how much is the variation in the radiated power as a function of direction away from the antenna. Mostly, the radiation patterns are evaluated in the far-field region [38]. Figure 7.7 illustrates the radiations of the Projected

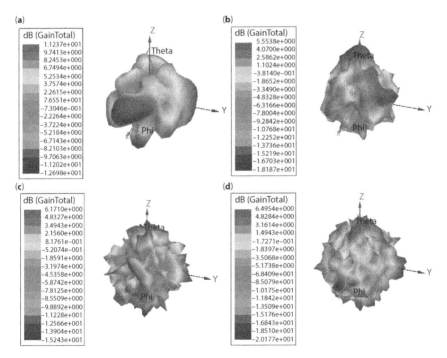

Figure 7.5 3-Dimensional gain plots at (a) 5.52 GHz (b) 11.36 GHz (c) 15.76 GHz (d) 18.05 GHz.

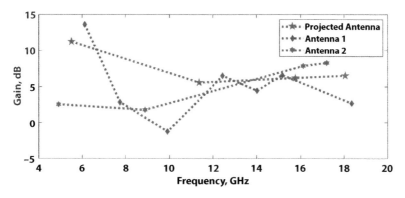

Figure 7.6 Comparison of gain values of Antenna 1, Antenna 2, and Projected Antenna.

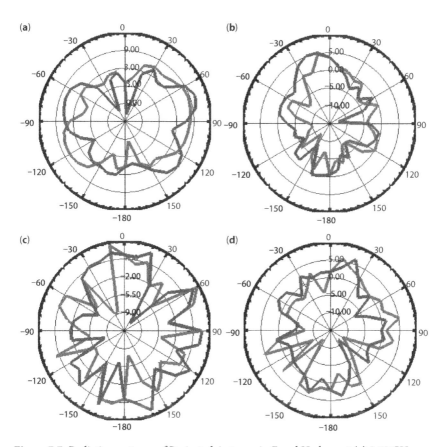

Figure 7.7 Radiation patterns of Projected Antenna in E and H planesat (a) 5.52 GHz (b) 11.36 GHz (c) 15.76 GHz (d) 18.05 GHz.

Antenna at the investigated frequencies. Such patterns are consistent and are stable. Examined 2-Dimensional patterns are nearly bidirectional/omnidirectional.

7.5.6   Surface Current Distribution

The surface current distribution provides a better understanding of the performance and radiation behavior of the antenna at the operating frequencies [34]. The simulated current distribution on the radiating patch is shown in Figure 7.8. Surface currents are highly concentrated at the lower half of the conductor as well as at the feed [38]. The current distribution decreases from the feed to the upper part of the conductor. At the upper portion, it is almost uniform. This way of current distribution is responsible for multiband behavior.

7.5.7   Axial Ratio

This parameter gives information about the nature of polarization of the designed antenna. Its value is calculated from the ellipse by taking the ratio between its minor and major axis [9]. If both the minor and major axis is the same and identical, then the antenna is treated as circularly polarized. At that time, its value is 0 dB. The term polarization illustrates the path opted by an electric field vector with respect to time [10]. Figure 7.9 reveals the plot of the axial ratio against the frequency of the Projected Antenna.

7.5.8   Group Delay

It is an essential metric that informs about the distortion/dispersion associated with the transmitted signal. In other words, it indicates phase distortion [9]. In the beginning, the almost same value of delay is observed. But, at some frequencies, the delay gets decreased while at other frequencies, the value gets increased. Figure 7.10 reveals the plot of the axial ratio against the frequency of the Projected Antenna.

7.6   Parametric Analysis

To understand deeply the physical behavior of Projected Antenna, parametric evaluation is carried out using Ansoft HFSS [34]. Therefore, for

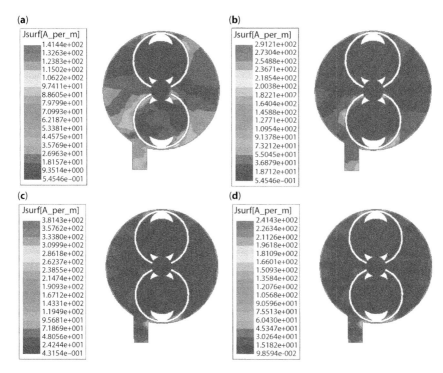

Figure 7.8 Current distribution of Projected Antenna in E and H planesat (a) 5.52 GHz (b) 11.36 GHz (c) 15.76 GHz (d) 18.05 GHz.

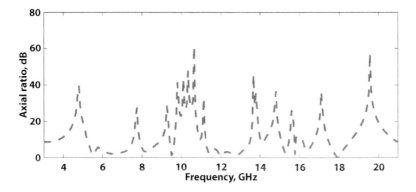

Figure 7.9 Axial ratio versus frequency of Projected Antenna.

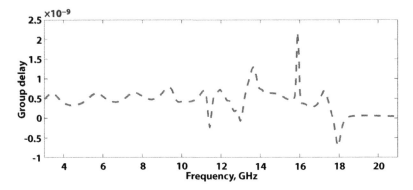

Figure 7.10 Group delay versus frequency of Projected Antenna.

such purpose, a few geometrical parameters are varied and the change in results is examined properly. The parameters selected for analysis are:- 'R_p', 'F_w', and 'L_{pG}'. Additionally, the influence of using different substrate materials is investigated.

7.6.1 Effect of Parameter 'R_p'

When the radius 'R_p' is varied, the whole design gets changed proportionately. The significant difference in operating frequency bands is examined by varying 'R_p' in the range of 11 to 13 mm. This parameter produces a significant influence on the radiation characteristics. By changing the value of the radius, all observations are noted down. This analysis is useful for getting the optimum value with superior results. Figure 7.11 reveals the S_{11} plot obtained by varying 'R_p' in the range of 11 to 13 mm.

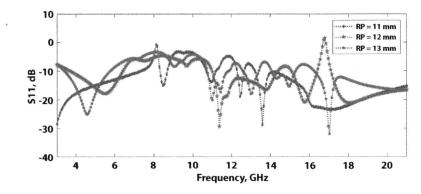

Figure 7.11 S11 characteristics with 'R_p' = 11, 12, and 13mm.

7.6.2   Effect of Parameter 'F$_w$'

The parameter 'F$_w$' is varied by an incremental factor of 0.5. Three values are taken in the range of 2.5 to 3.5 mm. There is a slight difference in the functional frequency bands. It is found that the selected value of 'F$_w$' reports that the aforementioned wideband antenna is suitable for wireless services. During analysis, the optimum value is 3 mm, as it provides multi-functionality and good gain. Figure 7.12 depicts the S$_{11}$ plot obtained by varying 'F$_w$'in the range of 2.5 to 3.5 mm.

7.6.3   Effect of Parameter 'L$_{PG}$'

The effect of ground plane dimension 'L$_{PG}$' is examined by assuming three values between 28 to 30 mm. The simulated results are examined after altering the value of 'L$_{PG}$'. From all the results, it is depicted that the optimal value of 'L$_{PG}$' is responsible for improved bandwidth and good impedance matching. A minimal change in S$_{11}$ characteristics is observed. Figure 7.13 shows the S$_{11}$ characteristics of Projected Antenna by varying 'L$_{PG}$' from 28 to 30 mm with an increment of 1 mm.

7.6.4   Effect of Different Substrate Materials

The selection of relevant substrate material is important as it greatly affects the resonating behavior of the device [44]. In this analysis, three types of substrate materials are considered: FR4, Glass, and Arlon AD410(tm). The simulation tool employed is kept the same for this evaluation. Dimensions

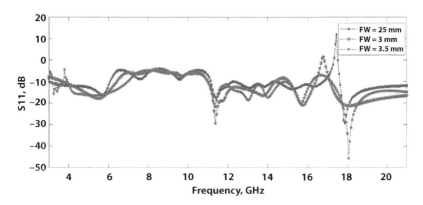

Figure 7.12 S$_{11}$ characteristics with 'F$_w$' = 2.5, 3, and 3.5 mm.

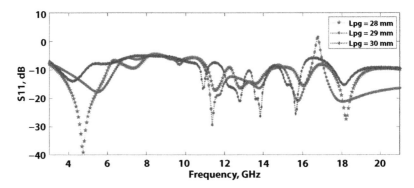

Figure 7.13 S11 characteristics with 'L$_{PG}$' = 28, 29, and 30 mm.

of the substrate get changed according to the type of material. As the design frequency is kept the same, but with a change in ε_r, the size of the antenna gets changes. Three designs are created based on the value of ε_r. In all the designs, the shape of the patch is the same but of a different size. In this work, it is found that by using different substrate materials, multi-band characteristics are observed. The only difference examined is that the value of gain is less than the fundamental value at a few operational frequencies. In other cases, bandwidth is less than the reported bandwidth of the Projected Antenna. A comparison of S11 characteristics obtained with three different substrate materials is shown in Figure 7.14.

For a fair comparison, the outcomes of the Projected Antenna are correlated with the outcomes of a few selected antennas that are already designed for the same applications. The comparison is illustrated in Table 7.2. Factors considered for differentiation are material type/volume,

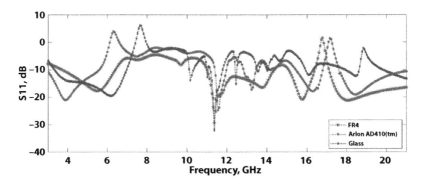

Figure 7.14 S$_{11}$ characteristics obtained with FR4, Glass and Arlon AD410 (tm).

Table 7.2 Comparison of Projected Antenna with few published multi-band antennas.

Ref. no.	Material/size (mm³)	Fundamental frequencies (GHz)	Description of Antenna
[20]	Glass PTFE/60 x 50 x 1.6	2.8/3.2/5.0	Microstrip antenna
[21]	FR4/50 x 60 x 1	0.9/2.4/3.8/5.1	Monopole antenna
[27]	FR4/70 x 70 x 0.8	2.45/5.51	Monopole antenna
[39]	FR4/88 x 108 x 1.6	2.0/3.5/4.9/6.5	X-shaped fractal antenna
[40]	FR4/120 x 87 x 1.6	0.36/1.32/5.50	H-fractal antenna
[41]	FR4/59 x 90 x 1.6	2.17/4.47/5.6	Fractal rhombic monopole antenna
[42]	FR4/56 x 59 x 1.6	1.57/2.66/3.63	Flower shaped monopole antenna
[43]	FR4/36 x 32 x 1.25	3.2/5.3/7.2/8.3/8.8	Wheel shaped fractal antenna
Projected Antenna	**FR4/32 x 32 x 1.6**	**5.52/11.36/15.76/18.05**	**Circular-Shaped patch antenna**

working frequencies, and structure type. Most of the published antennas are designed with FR4 epoxy material because it is easily available in the market and is cheap. From the comparison table, it is found that the Projected Antenna exhibiting compact structure with a unique shape, high gain, and good electromagnetic properties is claimed as a strong candidate for multiband wireless applications in comparison to other antennas found in the literature.

7.7 Summary

Wireless technology is preferred in several types of modern electronics as it offers greater mobility and convenience. In this chapter, a modified circular-shaped microstrip patch antenna is implemented for multiband applications by incorporating two identical patterns. The structure is backed by a partial ground plane. The step-by-step generation process is discussed and at each stage, the design is simulated and analyzed. The substrate size chosen for all stages is 32 x 32 mm^2. Results obtained at all the stages are compared. Projected Antenna, which is a combination of two similar patterns yields promising results for the aforesaid applications. The proposed prototype illustrates the multiband functionality by offering resonances at different wireless bands. It yields an operational bandwidth of 3.66-6.43 GHz, 11.00-14.38 GHz, 15.04-16.42 GHz, and 17.09 > 21 GHz at the respective working frequencies. Gain appeared at the associated frequencies is 11.23, 5.55, 6.17, and 6.49 dB. The corresponding VSWR values are 1.16, 1.28, 1.73, and 1.20, i.e., all the values are < 2. The observed multiple resonances are due to similar patterns utilized in the design and location of the feed. Current distribution and radiation pattern plots are presented at the claimed frequencies for deep analysis. A parametric study is conducted that effectively describes the significance of important design parameters. In addition, for proper understanding, the effect of three distinct substrate materials is demonstrated.

References

1. Jindal, S., Sivia, J.S., Bindra, H.S., Hybrid Fractal Antenna Using Meander and Minkowski Curves for Wireless Applications, *Wirel. Pers. Commun.,* 109, 4, 1471-1490, 2019.

2. Chowdhury, B.B., De, R., Bhowmik, M., A Novel Design for Circular Patch Fractal Antenna for Multiband Applications, *3rd International Conference on Signal Processing and Integrated Networks (SPIN-2016)*, pp. 449-453, 11-12 Feb., 2016, Noida.

3. Choukiker, Y.K., Sharma, S.K., Behera, S.K., Hybrid Fractal Shape Planer Monopole Antenna Covering Multiband Wireless Communications with MIMO Implementation for Handheld Mobile Devices, *IEEE T Antenna Propag.*, 62, 3,1483-1488, 2014.

4. Bangi, I.K., Sivia, J.S., Moore, Minkowski and Koch Curves Based Hybrid Fractal Antenna for Multiband Applications,*Wirel. Pers. Commun.*, 108, 4, 2435-2448, 2019.

5. Ahmad, A., Arshad, F., Naqvi, S.I., Amin, Y., Tenhunen, H., Loo, J., Flexible and Compact Spiral-Shaped Frequency Reconfigurable Antenna for Wireless Applications, *IETE J. Res.*, 66, 1, 22-29, 2018.

6. Varamini,G., Keshtkar, A., Daryasafar, N., Moghadasi, M.N, Microstrip Sierpinski fractal carpet for slot antenna with metamaterial loads for dual-band for wireless applications, *Int J Electron Commun.*, 84, 93-99, 2018.

7. Kaur, K., Sivia, J.S., A Compact Hybrid Multiband Antenna for Wireless Applications, *Wirel. Pers. Commun.*, 97, 4, 5917-5927, 2017.

8. Sharma, N., Bhatia, S.S., Double split labyrinth resonator-based CPW-fed hybrid fractal antennas for PCS/UMTS/WLAN/Wi-Max applications, *J Electromagnet Wave*, 33, 18, 2476-2498, 2019.

9. Balanis, C.A. *Antenna theory: Analysis and design*, 3rd ed., John Wiley & Sons, London, 2005.

10. Prajapati, P.R., Murthy, G.G.K., Patnaik, A., Kartikeyan, M.V., Design and testing of a compact circularly polarised microstrip antenna with fractal defected ground structure for L-band applications, *IET Microw. Antennas Propag.*, 9, 11, 1179-1185, 2015.

11. Parshad, K.D., *Antenna and Wave Propagation*, 3rd ed. Satya Parkashan, New Delhi, 2005.

12. Ozkaya, U., Seyfi, L., Dimension Optimization of Microstrip Patch Antenna in X/Ku Band via Artificial Neural Network, *Procedia Soc Behav Sci.*, 2520-2526, 2015.

13. Murugan, S., Rajamani, V., Design of Wideband Circularly Polarized Capacitive fed Microstrip Antenna, *Procedia Eng.*, 30, 372-379, 2012.

14. Khandelwal, M.K., Kanaujia, B.K., Dwari, S., Kumar, S., Gautam, A.K., Analysis and design of wideband micro strip-line-fed antenna with defected ground structure, *Int J Electron Commun.*, 68, 10, 951-957, 2014.

15. Deshmukh, A.A., Jain, S., Bagaria, T., Parekh, I., Mahale, S., Multi-band Slot Cut E-shaped Sectoral Microstrip Antennas, *Procedia Comput. Sci.*, 49, 319-326, 2015.

16. Krishna, R.V.S.R., Kumar, R., Design and investigations of a microstrip fed open V-shape slot antenna for wideband dual slant polarization, *Int. J. Eng. Sci. Technol.*, 18, 4, 513-523, 2015.

17. Nasimuddin, Chen, Z.N., Qing, X., Slotted Microstrip Antennas for Circular Polarization with Compact Size, *IEEE Antenn Propag M.*, 55, 2, 124-137, 2013.

18. Ghosh, A., Chakraborty, S., Chattopadhyay, S., Nandi, A., Basu, B., Rectangular microstrip antenna with dumbbell shaped defected ground structure for improved cross polarised radiation in wide elevation angle and its theoretical analysis, *IET Microw. Antennas Propag.*, 10, 1, 68-78, 2015.

19. Shanmuganantham, T., Raghavan, S., Design of a compact broadband microstrip patch antenna with probe feeding for wireless applications, *Int J Electron Commun.*, 63, 8, 653-659, 2009.

20. Mandal, K., Sarkar, P.P., A compact high gain microstrip antenna for wireless applications, *Int J Electron Commun.*, 67, 12, 1010-1014, 2013.

21. Karia, D.C., Goswami, S.A., Dhengale, B., A compact wideband antenna for wireless applications using rectangular ring, *1st International Conference on Next Generation Computing Technologies (NGCT)*, pp. 560-562, Dehradun, 2015.

22. Khanna, A., Srivastava, D.K., Saini, J.P., Bandwidth enhancement of modified square fractal microstrip patch antenna using gap-coupling, *Int. J. Eng. Sci. Technol.*, 18, 2, 286-293, 2015.

23. Jo, S., Choi, H., Shin, B., Oh, S., Lee, J., A CPW-Fed Rectangular Ring Monopole Antenna for WLAN Applications, *Int J Antenn Propag.*, 2014, 1, 1-6, 2014.

24. Vivek, R., Sreenath, S., Vinesh, P.V., Vasudevan, K., Coplanar Waveguide (CPW)-Fed Compact Dual Band Antenna for 2.5/5.7 GHz Applications, *Prog. Electromagn. Res. M.*, 74, 51-59, 2018.

25. Rahimi, M., Keshtkar, A., Zarrabi, F.B., Ahmadian, R., Design of compact patch antenna based on zeroth-order resonator for wireless and GSM applications with dual polarization, *Int J Electron Commun.*, 69, 1, 163-168, 2015.

26. Bhatia, S.S., Sivia, J.S., A Novel Design of Circular Monopole Antenna for Wireless Applications, *Wirel. Pers. Commun.*, 91, 3, 1153-1161, 2016.

27. Kumar, S.A., Dileepan, D., Design and development of CPW fed monopole antenna at 2.45 GHz and 5.5 GHz for wireless applications, *Alex. Eng. J.*, 56, 2, 231-234, 2017.

28. Shakib, M.N., Moghavvemi, M., Mahadi, W.N.L., Design of a Compact Tuning Fork-Shaped Notched Ultrawideband Antenna for Wireless Communication Application, *Sci. World J.*, Article ID 874241, 2014.

29. Jung, J., Lee, H., Lim, Y., Broadband flexible meander line antenna with vertical lines, *Microw. Opt. Technol. Lett.*, 49, 8, 1984-1987, 2007.

30. Kumar, R., Surushe G., Design of microstrip-fed printed UWB diversity antenna with tee crossed shaped structure, *Int. J. Eng. Sci. Technol.*, 19, 2, 946-955, 2016.

31. Raval, F., Kosta Y.P., Joshi, H., Reduced size patch antenna using complementary split ring resonator as defected ground plane, *Int J Electron Commun.*, 69, 8, 1126-1133, 2015.

32. Wang C.J., Li, S.C., Sun T.L., Lin, C.M., A wideband stepped-impedance open-slot antenna with end-fire directional radiation characteristics, *Int J Electron Commun.*, 67, 3, 175-181, 2013.

33. So, K.K., Wong, H., Luk, K.M., Chan, C.H., Miniaturized Circularly Polarized Patch Antenna with Low Back Radiation for GPS Satellite Communications, *IEEE T Antenna Propag.*, 63, 12, 5934-5938, 2015.

34. Bangi I.K., Sivia, J.S., Minkowski and Hilbert Curves Based Hybrid Fractal Antenna for Wireless Applications, *Int J Electron Commun.*, 85, 159-168, 2018.

35. Kaur, M., Sivia, J.S., Giuseppe Peano and Cantor set fractals based miniaturized hybrid fractal antenna for biomedical applications using artificial neural network and firefly algorithm, *Int J RF Microw C E.*, 30, 1, 1-11, 2019.

36. Singh, J., Sharma, N., A Comparison of Minkowski, Compact Multiband and Microstrip Fractals with Meander Fractals Antenna, *Int. J. Comput. Appl.*, 154, 4, 23-25, 2016.

37. Kaur, M., Sivia, J.S., Minkowski, Giuseppe Peano and Koch Curves based Design of Compact Hybrid Fractal Antenna for Biomedical Applications using ANN and PSO, *Int J Electron Commun.*, 99, 14-24, 2019.

38. Kaur, M., Sivia, J.S., ANN and FA Based Design of Hybrid Fractal Antenna for ISM Band Applications, *Prog. Electromagn. Res. C*, 98, 127-140, 2020.

39. Gupta, A., Joshi, H.D., Khanna,R., An X-shaped fractal antenna with DGS for multiband applications, *Int J Microw Wirel Technol.*, 9, 5, 1075-1083, 2017.

40. Weng, W.C., Hung, C.L., An H-Fractal Antenna for Multiband Applications,*IEEE Antennas Wire. Propag. Lett.*, 13, 1705-1708, 2014.

41. Mahatthanajatuphat, C., Saleekaw, S., Akkaraekthalin, P., A Rhombic patch monopole antenna with modified Minkowski fractal geometry for UMTS, WLAN and mobile WiMAX application, *Prog. Electromagn. Res.*, 89, 57-74, 2009.

42. Ullah, S., Faisal, F., Ahmad, A., Ali, U., Tahir F.A., Flint, J.A., Design and analysis of a novel tri-band flower-shaped planar antenna for GPS and WiMAX applications, *J Electromagnet Wave.*, 31, 9, 927-940, 2017.

43. Gupta, M., Mathur, V., Wheel shaped modified fractal antenna realization for wireless commucations, *Int J Electron Commun.*, 79, 257-266, 2017.

44. Kaur, M., Sivia, J.S., ANN-based Design of Hybrid Fractal Antenna for Biomedical Applications, *Int. J. Electron.*, 106, 8, 1184-1199, 2019.

8

Reconfigurable Antenna for Cognitive Radio System

Dr. Swapnil Srivastava[1*], Vinay Singh[1] and Dr. Sanjeev Kumar Gupta[2]

Electronics Department, Dr. Shakuntala Misra National Rehabilitation University, Lucknow, India
Electronics Department, AISECT University, Bhopal, India

Abstract

The most favorable technique to competently exploit the RF spectrum is Cognitive radio. The size of the antenna starts to become challenging when the DVB-H band is targeted almost 470 to 862 MHz; for miniaturization technique Metamaterial concept is widely used. The antenna attained multiband operation by stuffing with a metamaterial, the dispersion relations of the unit cell control these bands. In this paper, the design of four switches radiating buildings, integrating wideband plus narrowband antennas for Cognitive Radio has been proposed. This designed structure consists of a UWB antenna for spectrum detection as well as double narrowband antennas designed for wireless transmission taking place on the same FR4 substrate.

Keywords: Frequency reconfigurable antennas, cognitive radio system, ultra-wideband

8.1 Introduction

The requirement for a higher data rate is increasing as a result of the conversion from speech only multimedia transmission. The use of dynamic spectrum access has become a must because of the present fixed frequency facility. Cognitive radio (CR) is a well-thought-out and one of the most promising and innovative DSA techniques due to two exclusive parameters, cognitive capability as well as reconfigurability [1]. IEEE Standard has

---

Corresponding author: swapnil157@gmail.com

Prashant Ranjan, Dharmendra Kumar Jhariya, Manoj Gupta, Krishna Kumar, and Pradeep Kumar (eds.) *Next-Generation Antennas: Advances and Challenges*, (143–154) © 2021 Scrivener Publishing LLC

defined an antenna as a means for radiating or receiving radio waves. The antenna converts energy from electromagnetic to electrical and from electrical to electromagnetic. Frequency reconfigurable antennas have gained a lot of attention by adapting their properties to attain selectivity in gain, frequency, bandwidth, and polarization due to the development of innovative communication systems.

A reconfigurable antenna is capable of adjusting its radiation properties and frequency dynamically in a controlled and reversible method. Reorganization of RF currents above the antenna surface that produces alterable changes above its specifications is possible because frequency reconfigurable antenna integrates interior devices such as varactors, RF switches, mechanical actuators, and tunable materials [1]. The variance between the smart antenna and reconfigurable antenna is that the reconfiguration mechanism lies inside the antenna rather than in an external beam forming network. In a dynamic scenario to satisfy the changing operating requirements, the reconfiguration capability of reconfigurable antennas enhances the performance of the antenna. A reconfigurable antenna uses various ways to attain the change in parameters. The technique is based on expending switches such as diodes, GaAs FETs, or MEMS switches [1, 2].

The other hopeful approaches to surpass the vast biasing problems of electronic switches contains the use of optical switches and mechanical arrangement change. Due to the conversion from the audio transmission to multimedia type uses, the requirement for a higher data rate is growing. CR has two exceptional properties – cognitive capability and reconfigurability, hence it is considered one of the most capable and innovative DSA techniques. A lot of researches have been completed on CR.

8.2   Antenna

A metallic device or a transducer that radiates and receives EM waves is called an Antenna. One more definition introduces the antenna as the conversion from directed Electromagnetic signal and an empty-space Electromagnetic signal also between free-space EM signal and a guided EM signal. This technique is defined by universal communication between a transmitting and receiving antenna. Transmitting cable has the system of coaxial line/waveguide. When the coaxial cable is connected to the transmitter it radiates an RF signal which is directed by the uniform part of the line. Transverse electromagnetic signal with minimal loss transformed to an indicator which is amplified and applied to the antenna and when the

Rx antenna is taken into consideration, the Tx cable is connected to Rx that collects ACs, shown in Figure 8.1 given below [6].

Antenna radiation produces a definite length of the current element and a time-changing current/acceleration/deceleration of charge happens as defined in equation 8.1.

$$l.\frac{dl}{dt} = l.qi.\frac{dv}{dt}\left(A.\frac{m}{s}\right)$$

(8.1)

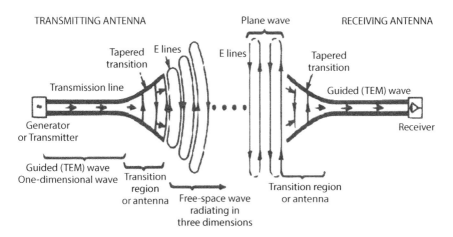

Figure 8.1 Typical structure of the antenna [6].

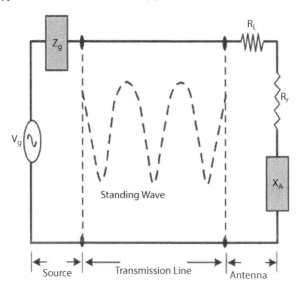

Figure 8.2 Antenna circuit as a whole structure [7].

Where, time alternating current in A/s, Charge in coulombs/m, and length in the meter.

Space in-between the cables of the Tx line is a portion of wavelength, so as the transition curve of an antenna open-up in order of wavelength the more the wave tends to radiate and launch in free-space. The propagation is vertical to acceleration and power is proportionate to the square of both portions of the equation mentioned above with the complete circuit of an antenna, shown in Figure 8.2 [7].

8.3   Antenna Reconfigurations

Classification of a reconfigurable antenna is done based on the antenna factor that is adjusted with dynamism, normally the operation frequency, the pattern of radiation and polarization [5]. In an antenna, reconfigurability is the capacity to change the fundamental operating features of a radiator through electrical, mechanical, or other means. Reconfigurable antennas are divided into four groups based on the properties of the reconfiguration shown in Figure 8.3, which is given below.

8.4   Uses and Drawbacks of Reconfigurable Antenna

For various applications, it has a multiband antenna in a single terminal. It is also miniature and easy to assimilate with swapping devices and control circuits. The structure of frequency reconfigurable antenna depends upon the equilibrium of trade-offs, in comparison to the static antenna due to its

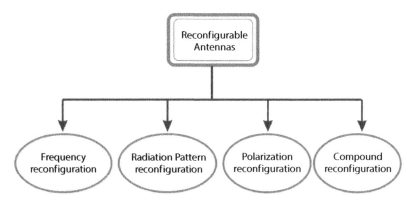

Figure 8.3 Reconfigurable antennas.

little evolving time. The technology of frequency reconfigurable antenna depends largely on RF switch technology which is not advanced enough; moreover, it has increased complexity and price with less efficiency.

8.5   Spectrum Access and Cognitive Radio

In today's scenario the need for wireless connectivity, as well as current crowding of permitted/restricted spectra, necessitate a novel communication model to make use of the current spectrum in enhanced ways. Spectrum circulation is grounded on allocating a certain frequency to definite provision. According to FCC Spectrum Policy Task Force, enormous temporal and geographic differences are present in the application of assigned spectrum in consumption range varying between 15% - 85% below the band of 3GHz. In the band above 3GHz, they do not function properly [3]. The static governing and licensing procedure is responsible for this inadequacy, which allots total rights to a frequency band to the main operator. The method makes tough the reuse of the frequency range as soon as they are assigned if this valuable resource is not used properly. The solution to this ineffectiveness which is useful in the ISM 2.4GHz, U-NII 5GHz-6GHz and microwave 57GHz-64GHz is to mark spectra existing on a restricted base [3].

8.6   Cognitive Radio

The main technology that offers the capacity to segment the wireless channel to the permitted operators in an adaptable manner is CR technology. CRs are predicted as skilled to offer a higher frequency range to mobile users through various wireless structures and energetic spectrum contact methods [3]. The spectrum must be shared with registered users by each cognitive radio without interfering and meet the varied worth of service utility. CR has to be reconfigurable, cognitive as well as organized by itself to fulfill functions of spectrum sharing, spectrum mobility, spectrum and spectrum decision. The ability of CR to detect the spectrum and hovels [3] is known as white spaces that aren't used by the permitted operators.

8.7   Spectrum Sensing and Allocation

In spectrum sensing CR scans the spectrum and identifies the idle portions of the spectrum. CR is known for its inside condition as well as outside atmosphere; it decides the perfect frequency range, then initiates the communication. The methods for spectrum sensing and communication are possible with the use of parallel and combined sensing.

Cognitive Radio design-parallel sensing is shown in Figure 8.4 given below [4].

In combined sensing, only one channel is used for spectrum sensing as well as communication, both, that is shown in Figure 8.5 given below [4].

For this operation, two antenna systems are designated. One antenna is Omni-directional and the second one is wideband. The first antenna nourishes the receiver, skilled in rough as well as fine spectrum sensing over a wide-ranging bandwidth and the other antenna is an indicator and nourishes a frequency responsive forward-facing end that can be adjusted to the designated band [4]. The solution for this is that the particular wideband antenna nourishes the spectrum sensing module as well as the frequency responsive forward-facing end. In the second method the spectrum recognizing and radio reconfiguration is done when the communication bond value falls inferior to defined standards. Two benchmarks are considered in this process. Link value decreasing below the first defined benchmark initiates spectrum sensing, therefore the improved system alignment is recognized which will complete the link superiority requirements. When the worth falls lower than a second lower benchmark the arrangement is reconfigured.

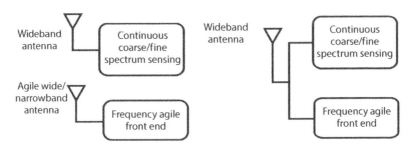

Figure 8.4 CR construction with parallel sensing.

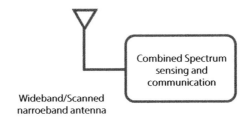

Figure 8.5 CR construction with combined sensing.

8.8   Results and Discussion

We have used a simulator of high frequency, CST 2017 built on a prede-termined time difference domain system and Reconfigurability of antenna frequency is achieved by implementing the switches from S1 to S4. All switches are linked with base design architecture. We are designing an antenna with four switches S1, S2, S3, S4 based antenna architecture as shown in Figures 8.6 and 8.7 given below.

In this proposed architecture the system is reconfigured by shifting its cur-rent route over the practice of switches from S1 to S4, respectively. Switches from S1 to S4 are used to attach-detach the metallic patches imprinted on the substrate. This architecture is designed with the help of the CST EDA tool. After examination, it is found that UWB exhibits the impedance BW of 24% from 0 to 12GHz for $|S_{11}| < -10$ dB. This antenna shows the resonance behavior at resonating frequencies (measured) 1.1, 2.81, 3.83, 4.94, 5.72,

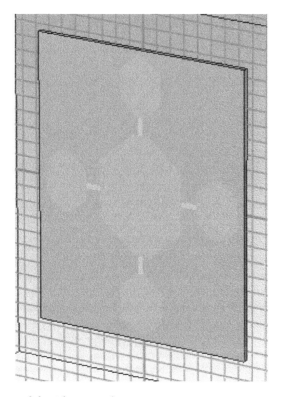

Figure 8.6 Base module without switch.

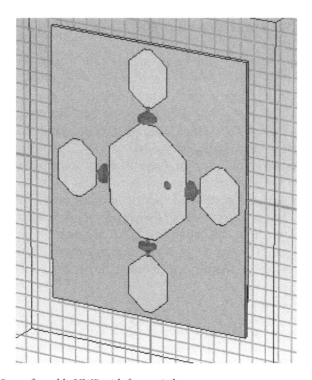

Figure 8.7 Reconfigurable UWB with four switch.

and 7.2 GHz. The series of the equation has been developed after inspecting the vector current distribution at resonating frequencies. The lower edge frequency of this antenna depends on the major axis radius of the elliptical patch. It has been found that the higher-order modes are generated at a higher frequency. Due to higher-order modes, the distorted pattern has been observed at frequency 3.83 GHz. In the second part of this chapter CWSA, the tuning element has been analyzed. In the evolution of this antenna, it has been examined that the slots and notch critically affect the impedance matching in the entire frequency span. Table 8.1 shows the different operation of switches and relative bandwidth at different frequency range as compared to the output achieved in the base paper [1].

In the base paper when all switches are off and the frequency range is 5 to 7 GHz, the relative bandwidth is 33.8%; when two switches are on, one switch is off and the frequency range is 4.3 to 6GHz & 7 to 9GHz, the relative bandwidth is 33.4% and 25.2%; when all three switches are on and the frequency range is 4 to 5.1GHz & 7 to 9GHz, the relative bandwidth is 26% and 25.2%.

Table 8.1 Comparison of impedance bandwidth for the three switching.

Switching operations of base paper	Base paper		Switching operations of designed system	Designed system	
	Frequency range	Relative bandwidth %		Frequency range	Relative bandwidth %
OFF OFF OFF	5 to 7 GHz	33.8%	OFF OFF OFF OFF	2 GHz to 6 GHz	33.3%
ON ON OFF	4.3 to 6 GHz & 7 to 9 GHz	33.4% & 25.2%	ON ON ON OFF	2 GHz to 12 GHz	31.6%
ON ON ON	4 to 5.1 GHz & 7 to 9 GHz	26% & 25.2%	ON ON ON ON	0 to 12 GHz	24%

In the proposed system when all switches are off and the frequency range is 2 to 6GHz the relative bandwidth is 33.3%; when three switches are on, one switch is off and the frequency range is 2 to 12GHz the relative bandwidth is 31.6 %; and when all four switches are on and the frequency range is 0 to 12GHz the relative bandwidth is 24%. Initially, it is designed without a switch, then connects the diode as a switch. We connect switch in all four sides of base architecture. After designing we analyze different scenarios of operation, operate the switch on and switch off cases and analyze the operation and frequency of the antenna. The architecture of our designed Reconfigurable UWB antenna generates the best output of multi-frequency band and optimized bandwidth. This is designed in multi-switch-based architecture and operates in different scenarios for multifunction operation.

Figure 8.8 shows basic S parameter results when all switches are off and the frequency range is 2 to 6GHz, the relative bandwidth is 33.3%. This

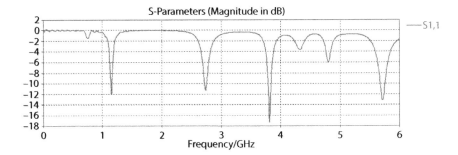

Figure 8.8 S-parameter when all switches are off.

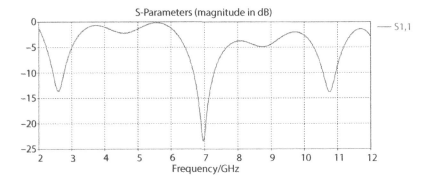

Figure 8.9 S-parameter when three switches are ON and one switch is OFF.

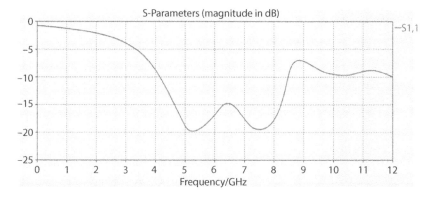

Figure 8.10 S-parameter when all switches are ON.

circuit is tested in different scenarios based on switch operation at different frequency bands between 2-12GHz as well as 0-12GHz.

Figure 8.9 shows the basic S parameter results when three switches are on, one switch is off and the frequency range is 2 to 12GHz, the relative bandwidth is 31.6%.

Figure 8.10 shows the S parameter result when all four switches are on and the frequency range is 0 to 12GHz, the relative bandwidth is 24%.

8.9   Conclusions

A cognitive radio system that is based on a frequency reconfigurable ultra-wideband antenna is designed to obtain a much higher bandwidth with improvement in other performance parameters as compared to the presently available system. The design has been successfully demonstrated to observe the behavior of the antenna, the parametric study of antennas has been carried out. After examination, it is found that four switches are used to reflect a better UWB case. The series of the equation has been developed after inspecting the vector current distribution at resonating frequencies. The lower edge frequency of this antenna depends on the major axis radius of the elliptical patch. It has been found that the higher-order modes are generated at a higher frequency. The frequency-domain parameters of these antennas have been perceived in terms of VSWR, S11 parameter, input impedance, and radiation pattern. USB exhibits the impedance BW of 24% from 0 to 12GHz for $|S\_11| < -10$ dB. This antenna shows the resonance behavior at resonating frequencies (measured) 1.1, 2.81, 3.83, 4.94, 5.72, and 7.2GHz.

References

1. Bakr, Mustafa S., *et al.* "Reconfigurable ultra-wide-band patch antenna: Cognitive radio." Advanced Materials and Processes for RF and THz Applications (IMWS-AMP), 2017 IEEE MTT-S International Microwave Workshop Series on. IEEE, 2017.
2. Tawk, Y., *et al.* "Reconfigurable front-end antennas for cognitive radio applications." *IET microwaves, antennas & propagation* 5.8 (2011): 985-992.
3. Al-Husseini, Mohammed, *et al.* "Reconfigurable microstrip antennas for cognitive radio." *Advancement in Microstrip Antennas with Recent Applications.* InTech, 2013.
4. Gardner, Peter, *et al.* "Reconfigurable antennas for cognitive radio: Requirements and potential design approaches." Wideband, Multiband Antennas, and Arrays for Defence or Civil Applications, 2008 Institution of Engineering and Technology Seminar on. IET, 2008.
5. Aizaz, Zainab, and Poonam Sinha. "A survey of cognitive radio reconfigurable antenna design and proposed design using genetic algorithm." Electrical, Electronics and Computer Science (SCEECS), 2016 IEEE Students' Conference on. IEEE, 2016.
6. Antenna Basics, McGraw- Hill Higher Education, Available: http:/highered.mcgraw-hill.com/sites/dl/free/0072321032/62577/ch02_011_056.pdf
7. Antenna Theory, Third Edition, Analysis And Design by Constantine A. Balanis, A John Wiley & Sons, INC., Publication, Chapter 1 "Antenna" Page 3

9

Ultra-Wideband Filtering Antenna: Advancement and Challenges

Prashant Ranjan[1]*, Krishna Kumar[2], Sachin Kumar Pal[3] and Rachna Shah[4]

[1]Department of ECE, University of Engineering and Management, Jaipur, India
[2]UJVN Ltd., Uttarakhand, India
[3]Bharat Sanchar Nigam Ltd., Guwahati, India
[4]National Informatics Centre, Dehradun, India

Abstract

Ultra-wideband (UWB) is a short-distance radio communication technology that allows wireless networking inside the digital office and home with reliably high data rates across multiple devices [1]. However, some narrowband communication systems have existed in the UWB band, such as WiMAX (worldwide interoperability for microwave access) at 3.6 GHz, WLAN (wireless local area network) at 5.2 GHz, and X–band for satellite communication systems. In order to minimize the effect of interferences between the UWB systems and narrowband systems, a great number of UWB antennas with filtering characteristics and having a notch band have been used [2]. Integration of filters with antennas plays an important role in designing of notch band UWB filtering antenna. In this chapter, a survey of UWB filtering antenna and UWB filtering antenna with notch band characteristics is presented.

Keywords: Ultra-wideband antenna, filtering antenna, high antenna gain, notch band, selectivity

9.1 Introduction

The filter and the antenna in the RF front-end system are two critical instruments. Typically, they are viewed as two distinct fields of study and are independently designed by various researchers. To provide better device

---

*Corresponding author: prashant4235@gmail.com

Prashant Ranjan, Dharmendra Kumar Jhariya, Manoj Gupta, Krishna Kumar, and Pradeep Kumar (eds.)
Next-Generation Antennas: Advances and Challenges, (155–164) © 2021 Scrivener Publishing LLC

efficiency in the application, the filter and antenna need to be integrated. There are several colourful UWB filters and antennas in the literature, but an increase in circuit size and loss would be caused by the interconnection between these two elements. Thus, if the filter and antenna can be combined to avoid using a transmission line for a link, it is very helpful. This would increase the efficiency of the device effectively, reduce the cost, and decrease the size of the circuit. A notched band filtering antenna is a very good example of a filtering antenna [13–20]. This chapter introduces the literature review of filtering antenna. Also discussed are their attractive features, advantages, and disadvantages in tabular format.

9.2   Ultra-Wideband Filtering Antenna

Mandal *et al.* [3] proposed a balanced UWB filtering antenna. Coupled–CPS is used for a wideband bandpass filter design. Suppress the undesired harmonic passband and created transmission zeros by using shunt open-stubs stripline. A differential UWB antenna cascaded with the filter in the feeding position of the antenna is presented. Because of the wide upper stopband nature of the integrated filter, the antenna with filter provides good higher band rejection of the signal. The bandwidth of the antenna plus filter is from 3.1 to 5.1 GHz. It covers the lower UWB band (3.1– 4.85 GHz). The filtering antenna achieves a good radiation pattern over the 3-5 GHz UWB band and suppresses the unwanted out-of-band signals. The characteristic of filtenna improves in the UWB band.

Tang *et al.* [4] proposed two planar ultra-wideband antennas. The first UWB antenna is without filtering characteristics and the second UWB antenna is combined with a filter to improve the realized gain performance. A single-wing element MMR filter is combined with the slot modified UWB antenna. A 61.7% size reduction from previous designs is presented with more than a 6 dB realized gain near 10 GHz. For further improvement in the UWB antenna, the filter is first designed and integrated with this compact filter with the UWB antenna. The integrated design holds sharp frequency cutoffs at both ends of the UWB passband and also possesses strong upper stopband attenuation. The filtering antenna achieves a further 2.12 dB increase in the broadside-realized gain near 10 GHz. It also improved the radiation pattern performances by achieving omnidirectional radiation patterns in the H–plane. Provide a bandwidth from 2.715 to 10.13 GHz with a fractional bandwidth of 115.45%. Single-wing filter antenna has a gain of 4.25 dBi.

Lee *et al.* [5] suggested the direct sequence UWB (DS-UWB) low band or multi-band orthogonal frequency division multiplexing (MB-OFDM) (3.1–5.2 GHz), an ultra-wideband antenna coupled with a filter. In the UWB bandpass filter, the filtering antenna presented consists of a trapezoidal radiating patch with microstrip feeding, an interdigital capacitor (IDC) and a dumbbell-shaped defective soil structure (DGS). The first move is to build a trapezoidal UWB antenna with a frequency range of 3.1-10.6 GHz. In the second stage, a filter was designed by the authors using DGS and IDC structures with a 3.1 to 5.2 GHz passband. Integrate the trapezoidal UWB antenna and the filter onto a single dielectric substrate in the final step. The filtering UWB antenna has a dimension of 30 x 41.2 mm^2. The filter limits the UWB antenna passband. The filtering antenna's bandwidth occupies the 3.1–5.2 GHz frequency spectrum. Radiation patterns look like a figure of eight and the UWB antenna filtering gain within the passband is 2.3 to 3 dBi.

Sahoo *et al.* [6] proposed a UWB integrated filter antenna (IFA) with improved selectivity. Electromagnetically, a circular radiating patch is coupled to a UWB filter. Losses are reduced since the filter and the antenna are not used to cascade with a matching thread. The performance of the filtenna is more than 90% in the desired band because of this. With antenna output retention in the UWB band, Filtenna can reject the unwanted out-of-band signal. The filtenna gain decreased after the upper cut-off frequency, but the gain is constant in the case of the antenna alone. The entire UWB band (3.1-10.6 GHz) is shielded by Filtenna and gives stronger rejection of the signal outside the band. The peak gain of the IFA varies between 3.0 to 4.0 dBi.

Wong *et al.* [7] proposed a UWB filter-antenna using via. Tapered shape radiating patch and ground plane is connected by via. Two resonant modes are provided due to the tapered shape structure. Filter–antenna performance is improved by adding four pins to shorting the antenna. A notch-band is generated in the UWB band by using a shunt LC-resonator. A pair of the short-circuited quarter-wavelength strip and a pair of half-wavelength pins are added to ground the antenna. It improves the antenna selectivity. The effect of the ground plane in the antenna is presented. In comparison with a small ground plane antenna, the large ground plane antenna has a larger gain. Current density is considered maximum along the edges of the tapered side, which provides wideband radiation.

Panda *et al.* [8] proposed a filtenna for UWB wireless communication system. Firstly, a wideband (2.65 – 8.52 GHz) monopole antenna fed with a microstrip transmission line is designed. Secondly, a planar UWB filter

Table 9.1 Summary of ultra-wideband filtering antenna.

Sl. no.	Author	Year	Covered bandwidth	Gain	Technique used	Limitation
1	Ming et al. [4]	2016	2.9 - 10.8 GHz	5 dBi	MMR	No flat passband gain
2	Jung et al. [5]	2008	3.1–5.2 GHz	4 dBi	Interdigital capacitor	The entire UWB band not covered
3	Anuj et al. [6]	2016	3.1–10.6 GHz	3-4d Bi	MMR with a triangular ring loaded stub	Minimum gain
4	Sai et al. [7]	2013	3.1–10.6 GHz	2-3 dBi	Using via and tapered shape	Minimum gain
5	Jyoti et al. [8]	2010	3.65 -10.16 GHz	NA	MMR	Poor selectivity
6	Young et al. [11]	2006	2.9 - 14.5 GHz	2 dBi	Half-bowtie radiating element and DGS	No sharp rejection at UWB band edges
7	Garcia et al. [12]	2015	3.1- 4.7 GHz	NA	DGS	The entire UWB band not covered
8	P. Ranjan et al. [22]	2017	3.1 - 11.01 GHz	9.8	Circularly-slotted-flower-shaped patch and MMR	No flat passband gain

is introduced in the place of microstrip transmission line feeding to make a filtering antenna. Due to the insertion of the UWB filter in place of the feeding line, the performance of the filtering antenna is improved and passband frequencies are bounded to the UWB frequency range. MMR filter is used to improve the upper passband frequency in the UWB antenna. Activated modes are overlapped suitably to achieve a UWB performance in the filtering antenna. The role of the filter is to provide an upper stopband at the higher edge of the UWB band which improves the selectivity of the antenna. The passband of filtenna is from 2.27 GHz to 10.33 GHz. Chen et al. [9] proposed a filter-antenna subsystem for UWB communications. A UWB antenna without a filter can't stop passband frequency at the higher edge of the UWB band (10.6 GHz). In this paper, an MMR based filter is combined with an antenna to design a UWB filtering antenna. The passband of the filter–antenna is from 3.3–10.4GHz. A quasi omnidirectional radiation pattern is obtained.

Yeo et al. [10] proposed a CPW-based half-ring-shaped (HRS) slot antenna for harmonic suppression applications. The antenna consists of symmetrically placed SIRs added to both ends of the HRS. Harmonics of the fundamental resonance frequency are suppressed by using two SIRs. Cho et al. [11] proposed a planar monopole antenna with a staircase shape UWB applications. The antenna consists of the staircase-shape, a half-bowtie conducting patch, and a DGS to achieve a UWB bandwidth. Measured bandwidth is from 2.9 – 14.5 GHz (11.6 GHz). U–shaped slot cut in the radiating patch behaves as a band rejected filter. A notch band is generated from 5 to 5.86 GHz. An omnidirectional radiation pattern is obtained. A summary of previous work on ultra-wideband filtering antenna is given in the following Table 9.1.

9.3   Ultra-Wideband Filtering Antenna with Notch Band Characteristic

Chung et al. [13] proposed a UWB antenna with a band-notch filter. A small strip bar and two microstrip patches of the same size are used to obtain UWB and notch band characteristics of the antenna. A notch band of 4.9 to 6.0 GHz is obtained between a passband of 3.1 to 13GHz. A small strip bar is used to control the band notch characteristic. Strip bar is behaving as a quarter-wave open-circuited stub. The antenna has observed omnidirectional radiation patterns in XZ–plane are.

Ahmadi et al. [14] proposed a miniaturized printed monopole UWB antenna with a notch band. The radiating element and its ground plane have a pentagonal shape and they are symmetrical. A tapered microstrip line is used for matching between the feed line and the input impedance of the antenna. V–shape or π–shaped slots are cut in a radiating patch to create band-notched characteristics. Notch band exists from 5.725 – 5.825 GHz for rejection of the WLAN band.

Karimian *et al.* [15] proposed two monopole UWB antennas with band-notched characteristics. A tree-shaped conducting patch with DGS is used to design the UWB antennas having FBW of more than 170% (2.81 – 9.2 GHz). Two different approaches are used to reject desired-frequency bands. In the first approach, reject the 5.1 – 5.9 GHz frequency band by using a T–shaped resonator. In the second approach, reject 3.4 – 3.6 GHz and 5 – 5.6 GHz of the frequency band by using two L-shaped slots in the patch. By controlling the length of the slots the band notched can be controlled. Antennas have omnidirectional radiation patterns in the XZ–plane, and monopole like patterns in the YZ–plane. Antenna gains up to 3 dB is achieved. In this paper undesired out-of-band gain performance is poor.

Azim *et al.* [16] proposed a planar antenna with dual notched bands for UWB applications. The antenna is designed by etching a rectangular patch on the top of the substrate and partial ground on the other side of the substrate. Due to strong coupling between the radiating element and partial ground plane, capable to achieve a wide frequency band from 2.9 to more than 11 GHz. A tri-arm slot below the patch is etched to achieve the two band notches at 3.5 and 5.5 GHz. Abdalla *et al.* [17] proposed a notch band ultra-wideband antenna. The design is based on switching techniques. The radiating element of the UWB antenna looks like a half-circle. Curvature shaped partial ground plan is used for impedance matching. Switched meandered slots etched on the radiation patch and in the microstrip feed line to achieve two-notch band frequencies. Four modes are used to control the switches ON and OFF. Two notch bands from 3.3 to 3.8 GHz and from 5.1 to 5.9 GHz are achieved. The antenna has a nearly circular radiation pattern which is better for UWB application.

Tang *et al.* [18] proposed a notch band UWB antenna. The antenna consists of two quasi-square radiating patches and an octagonal slot in the ground plane. Three notch bands are produced by employing CSRRs on the ground plane and SRRs on the radiating patch. Notch bands are filtered the frequency bands of WiMAX, WLAN, and ITU bands at 3.5 GHz, 5.5 GHz, and 8 GHz respectively. Impedance bandwidths from 2.55 to 10.86 GHz with three notch bands are achieved. Huang *et al.* [19] proposed a polarization diversity-based UWB antenna having two notch bands. Antenna contains two orthogonal differential pairs and a chamfer

Table 9.2 Summary of ultra-wideband filtering antenna with notch band characteristics.

Sl. no.	Author	Year	Notch band frequencies	Gain	Technique used	Limitation
1	Chung et al. [13]	2007	5.5 GHz	5	Small strip bar	Not bounded on UWB band
2	Ahmadi et al. [14]	2009	5.8 GHz	NA	Pentagonal shape and V-shaped slot	Maximum return loss at out of passbands
3	Karimian et al. [15]	2014	3.5 and 5.8 GHz	1.5 dBi	Sharp and sudden discontinuity in the feed line and T-shaped with via	Not bounded on UWB band
4	Azim et al. [16]	2014	3.5 and 5.5 GHz	Up to 6 dBi	Single tri-arm resonator	Not limited to UWB band
5	Mahmoud et al. [17]	2015	3.3 to 3.8 GHz and from 5.1 to 5.9 GHz	Up to 4.5 dB	Defected microstrip structure (DMS)	Not bounded on UWB band
6	Tang et al. [18]	2015	3.21–3.86, 4.93–6.13 and 7.87–8.89 GHz	Up to 4 dBi	Split-ring resonators (SRRs) and complementary SRRs	Poor selectivity
7	Huang et al. [19]	2015	3.3–3.7 GHz and 5.15–5.8 GHz	Up to 6 dBi	CSRR slots and open-circuited stubs	Poor return loss
8	Li et al. [20]	2016	3.35-3.55 GHz and 5.65-5.95 GHz	Up to 4.2 dB	Complementary splitting resonator (CSRR)	Maximum insertion loss at passbands

square slot to achieve polarization diversity over the UWB. The bandwidth of the antenna from 2.75 GHz to more than 11 GHz is obtained. Each radiator has a CSRR slot and an open stub, to obtain a dual band-notched at 3.5 and 5.5 GHz. Measured gain varies between 3 to 6 dBi within desired frequency bands.

Li *et al.* [20] proposed a filtenna for UWB–MIMO applications with good out-of-band characteristics. A BPF of three interdigital microstrip lines and shunt stubs are integrated into the feed line of the printed antenna to achieve a filtering characteristic and improve the selectivity. To obtain the dual notched band characteristics a CSRR is introduced into the patch. The measured impedance bandwidth is 3.1 GHz to 10.65 GHz with two notch bands from 3.35 GHz to 3.55 GHz and 5.65 GHz to 5.95 GHz. Mardani *et al.* [21] proposed a monopole antenna with a tunable notch band function. The antenna consists of a radiating patch, partial ground plane, and two inverted L–shaped stubs on the upper portion of the ground plane. Pins are used to connect the upper patch and stubs on the ground plane. Pins are acted as resonant elements to produce filtering function. The position of the notch bands can be controlled by lengths of L–shaped stubs. The bandwidth of the antenna is from 3 – 10.6 GHz with notch bands at 3.5 and 5.5 GHz. A summary of previous work on ultra-wideband filtering antenna with notch band characteristics is given in the following Table 9.2.

9.4   Conclusions

The review of literature in this chapter has concentrated largely on empirical observations of filtering antenna design for UWB applications. Various design techniques reported in the literature for UWB filtering antenna, UWB filtering antenna with notch band, and their design issues have been reviewed and discussed. Many researchers have explained multiple band creation, notch band creation, bandwidth enhancement technique based on stub loaded open-loop resonator, split-ring resonator, and defected ground structure, and these have been investigated in detail. Some authors have discussed gain enhancement, size reduction, omnidirectional type radiation pattern, efficiency improvement, power consumption, integration of filter with antenna technique based on multiple mode resonator, stepped impedance resonator, and meander line, etc. It is found that sharp selectivity, bandwidth enhancement, gain enhancement, size reduction, and interference avoidance are important issues related to UWB filtenna.

References

1. Walter Hirt, "Ultra-wideband radio technology: overview and future research," *Computer Communications*, vol. 26, pp. 46–52, 2003.
2. A. Djaiz, M. A. Habib, M. Nedil and T. A. Denidni, "Design of UWB filter-antenna with notched band at 5.8GHz," *2009 IEEE Antennas and Propagation Society International Symposium, Charleston, SC*, 2009, pp. 1-4.
3. M. K. Mandal and Z. N. Chen, "Compact wideband coplanar stripline bandpass filter with wide upper stopband and its application to antennas," in *IET Microwaves, Antennas & Propagation*, vol. 4, no. 12, pp. 2166-2171, December 2010.
4. M. C. Tang, T. Shi and R. W. Ziolkowski, "Planar Ultrawideband Antennas With Improved Realized Gain Performance," in *IEEE Transactions on Antennas and Propagation*, vol. 64, no. 1, pp. 61-69, Jan. 2016.
5. Jung N. Lee, Jong K. Park, and II H. Choi, "A Compact Filter-Combined Ultra-Wide Band Antenna for UWB Applications," *Microwave and Optical Technology Letters*, vol. 50, no. 11, pp. 2839-2845, November 2008.
6. Anuj Kumar Sahoo, Ravi Dutt Gupta, and Manoj Singh Parihar, "Highly Selective Integrated Filter Antenna for UWB Application," *Microwave and Optical Technology Letters*, vol. 59, no. 5, pp. 1032-1037, May 2017.
7. S. W. Wong, T. G. Huang, C. X. Mao, Z. N. Chen and Q. X. Chu, "Planar Filtering Ultra-Wideband (UWB) Antenna with Shorting Pins," in *IEEE Transactions on Antennas and Propagation*, vol. 61, no. 2, pp. 948-953, Feb. 2013.
8. J. R. Panda, P. Kakumanu and R. S. Kshetrimayum, "A wide-band monopole antenna in combination with a UWB microwave band-pass filter for application in UWB communication system," *2010 Annual IEEE India Conference (INDICON), Kolkata*, 2010, pp. 1-4.
9. Y. Chen and Y. Zhou, "Design of a filter-antenna subsystem for UWB communications," *2009 3rd IEEE International Symposium on Microwave, Antenna, Propagation and EMC Technologies for Wireless Communications, Beijing*, 2009, pp. 593-595.
10. Junho Yeo and Jong-Ig Lee, "Compact CPW-Fed Half-Ring-shaped Slot Antenna for Harmonic Suppression Applications," *Microwave and Optical Technology Letters*, vol. 57, no. 5, pp. 1102-1104, May 2015.
11. Young Jun Cho, Ki Hak Kim, Dong Hyuk Choi, Seung Sik Lee and Seong-Ook Park, "A miniature UWB planar monopole antenna with 5-GHz band-rejection filter and the time-domain characteristics," in *IEEE Transactions on Antennas and Propagation*, vol. 54, no. 5, pp. 1453-1460, May 2006.
12. I. J. Garcia Zuazola *et al.*, "Band-pass filter-like antenna validation in an ultra-wideband in-car wireless channel," in *IET Communications*, vol. 9, no. 4, pp. 532-540, 2015.

13. K. Chung, S. Hong and J. Choi, "Ultrawide-band printed monopole antenna with band-notch filter," in *IET Microwaves, Antennas & Propagation*, vol. 1, no. 2, pp. 518-522, April 2007.

14. B. Ahmadi and R. Faraji-Dana, "A miniaturised monopole antenna for ultra-wide band applications with band-notch filter," in *IET Microwaves, Antennas & Propagation*, vol. 3, no. 8, pp. 1224-1231, December 2009.

15. R. Karimian, H. Oraizi and S. Fakhte, "Design of a compact ultra-wide-band monopole antenna with band rejection characteristics," in *IET Microwaves, Antennas & Propagation*, vol. 8, no. 8, pp. 604-610, June 4, 2014.

16. R. Azim, M. T. Islam and A. T. Mobashsher, "Dual Band-Notch UWB Antenna With Single Tri-Arm Resonator," in *IEEE Antennas and Wireless Propagation Letters*, vol. 13, pp. 670-673, 2014.

17. M. A. Abdalla, A. A. Ibrahim and A. Boutejdar, "Resonator switching techniques for notched ultra-wideband antenna in wireless applications," in *IET Microwaves, Antennas & Propagation*, vol. 9, no. 13, pp. 1468-1477, 10 22 2015.

18. Z. Tang, R. Lian and Y. Yin, "Differential-fed UWB patch antenna with triple band-notched characteristics," in *Electronics Letters*, vol. 51, no. 22, pp. 1728-1730, 10 22 2015.

19. H. Huang, Y. Liu and S. Gong, "Uniplanar differentially driven UWB polarisation diversity antenna with band-notched characteristics," in *Electronics Letters*, vol. 51, no. 3, pp. 206-207, 2 5 2015.

20. W. T. Li, Y. Q. Hei, H. Subbaraman, X. W. Shi and R. T. Chen, "Novel Printed Filtenna With Dual Notches and Good Out-of-Band Characteristics for UWB-MIMO Applications," in *IEEE Microwave and Wireless Components Letters*, vol. 26, no. 10, pp. 765-767, Oct. 2016.

21. H. Mardani, C. Ghobadi and J. Nourinia, "A Simple Compact Monopole Antenna With Variable Single- and Double-Filtering Function for UWB Applications," in *IEEE Antennas and Wireless Propagation Letters*, vol. 9, pp. 1076-1079, 2010.

22. Prashant Ranjan, Saurabh Raj, Gaurav Upadhyay, V. S. Tripathi, S. Tripathi, "Circularly slotted flower shaped UWB filtering antenna with high peak gain," *International Journal of Electronics and Communications (AEU)*, Vol. 81, pp. 209-217, 2017.

10

UWB and Multiband Reconfigurable Antennas

Manish Sharma[1]*, Rajeev Kumar[1] and Preet Kaur[2]

[1]*Chitkara University Institute of Engineering and Technology, Chitkara University, Punjab, India*
[2]*JC Bose University of Science and Technology, YMCA, Faridabad, Haryana, India*

Abstract

Reconfigurable antennas are prominently used in multiband applications due to their capability of changing operating frequency, polarization, and radiation characteristics. Based on the surrounding conditions and need for more prominent wireless applications, they can modify the geometry and the behavior by implementing different reconfigurable mechanisms such as RF PIN diodes, Varactor diodes, Micro-Electromechanical System (MEMS) switch, Field Effect Transistors (FET), metamaterial and even liquid crystals. In Ultra-wideband (UWB) technology, the bandwidth coverage is 3.10GHz-10.60GHz. Already existing wireless systems such as Wireless Interoperability for Microwave Access (WiMAX), Wireless Local Area Network (WLAN), and satellite uplink/downlink bands do interfere with the operating UWB bandwidth. Hence, filters are used to eliminate these interfering bands and by using different switching methods, these interfering bands are reconfigured or controlled independently. On the other hand, there are reported multiband antennas that resonate for different centered frequencies such as Long Term Evaluation LTE700/2300/2600, Global Systems for Mobile Communication GSM850/900/1900, Universal Mobile Telecommunication System UMTS2100, Cellular System, Personal Communication System (PCS), Bluetooth, WiMAX: 2.30GHz-2.40GHz/2.496GHz-2.690GHZ/3.30GHz-3.80GHz, WLAN: 2.412GHz-2.484GHz/5.250GHz-5.350GHz/5.725GHz-5.850GHz, Global Navigation Satellite System (GLONASS), UWB, X, Ku, and K-band. In this chapter, reconfigurable methodology with interfering bands and multiband antennas are discussed.

Corresponding author: manishengineer1978@gmail.com

Prashant Ranjan, Dharmendra Kumar Jhariya, Manoj Gupta, Krishna Kumar, and Pradeep Kumar (eds.)
Next-Generation Antennas: Advances and Challenges, (165–184) © 2021 Scrivener Publishing LLC

Keywords: Reconfigurable, PIN, varactor, MEMS, FET, UWB, multiband

10.1   Introduction

Antennas play a major role in the modern communication system and RADAR applications. Also, antennas can be categorized into different groups such as dipole/monopole, loop, horn, reflector helical, frequency-independent, and microstrip antennas. Here, the discussion of reconfigurable antennas is related to patch antennas with mitigating interfering bands and offering multiband characteristics. Furthermore, these antennas are reconfigured using different switching techniques where PIN diodes, Varactor diodes, RF MEMS switch, etc., are used. According to the change in the communication environment, reconfigurable antennas also should be capable to alter the operating frequency, impedance bandwidth, polarization, and radiation patterns. Hence, a reconfigurable definition for any antenna system can be defined as the change in the elementary characteristics which includes the change in electrical/mechanical or by other means. Reconfigurable techniques that are discussed above are employed either in notched band antennas [1–12] or for circularly polarized antennas [13, 14] or antennas meant for multiband applications [15–40]. In a triangular patch antenna covering a bandwidth of 3.05GHz to 13.10GHz is band-notched by etching a triangular slot on the patch. Placement of PIN diode on the triangular slot achieves reconfigurability by switching ON (WLAN is notched) and OFF (without a notch) the HPND4005 RF PIN diode [1]. Also, in a circular slot antenna, reconfigurable characteristics are maintained by using two switches into capacitance loaded split-ring resonators [2]. Stepped impedance resonators (SIRs) are also used to notch interfering bands and deployment of switches on SIRs results in reconfigurable characteristics [3]. T-type impedance resonator and parallel stubs loaded resonator are reconfigured by using four switches providing different modes of the configuration of the band-notched antenna [5]. Also, using three PIN diodes in the parasitic structure printed above the ground plane is reconfigured to different three interfering bands (WLAN/ITS/ITU) [6]. Four switchable frequency modes are obtained in the UWB antenna with dual interfering bands which are reconfigured by placing PIN diodes within the structure [8]. In triple notched wideband antenna, three interfering bands are reconfigured by using PIN diodes and in triple notched super wideband antenna where notches are achieved by using T-type stub and C-shaped slots are reconfigured by properly placing PIN diodes [10, 11]. According to the need for applications, antennas can be also polarized and can be reconfigured [13, 14]. The optical biased switching method is used in a C-type radiator and these switching OFF/ON mode provides three different states of operation, namely linear polarization (LP), Right-Handed Circular Polarization (RHCP), and Left-Handed Circular Polarization (LHCP) [13]. Furthermore, a single antenna is designed to achieve multiband applications [15–36] and is reconfigured where different switching techniques such as PIN diodes, MEMS switch, and Varactor diodes are used. Multiband antenna (WiFi, WiMAX, 5G) is reconfigured by using shunt capacitors which are placed over the gaps between the patches [20]. PIN diodes form the connectivity in triple frequency on-demand applications which is generated by CPW-fed triangular patch [23]. In a multiband antenna which characteristics are obtained by etching trapezoidal slot on the patch and rectangular slots in the ground is reconfigured by placing PIN diode switches between rectangular slot and ground [24]. A Multiband antenna resonating for five (DCS, PCS, WiFi, WiMAX & WLAN)/three (UMTS, WiMAX & WLAN)/bands is achieved by using stubs/slots on the patch or by using dual folded slots on the patch [40, 41], detailed study achieving multiple reconfigurable is reported [42] which includes the reconfigurable configuration in frequency, radiation pattern & polarization. Also, an equivalent circuit model of the wideband antenna with notched band characteristics is also reported [43].

10.2   Need for Reconfigurable Antennas

Change in the communication environment requires the antenna to operate with the capability of altering its operating frequencies, impedance bandwidths, polarization, and also radiation pattern, which are the features for any antenna designed for wireless applications. This deployment of antennas raises challenges to both the designed antenna as well as the designer. It is known that a single element antenna is extensively used in portable devices with wireless capability which includes cellular phones, PDAs (Personal Digital Assistant), or laptops. These wireless antennas are generally microstrip patch antennas which may or may not work on multiple frequency bands. In such a case, generally, one of the antennae is used for transmission. This communication link established between transmitter (from portable devices) to the base station or any other access point forms the weakest link in bidirectional communication due to the consumption of power, size, and limitations of the costing involved in manufacturing these portable devices. Hence reconfigurability in such a scenario provides a better solution for the above-said problem which can be tuned to the antenna's operating frequency and can be utilized to change operating bands, notch out interfering bands, or maybe tune the antenna to the change in a new environment. One of the best methods of saving power consumption is to redirect the radiation pattern to the access point and use the lower power of consumption which is achieved by reconfiguring the radiation pattern.

Classification of reconfigurable can be majorly divided into four groups, which are discussed below:

a) Shifting in the resonant frequency of the antenna by altering the structure using RF switches and thus frequency reconfigurable characteristics can be obtained

b) Reconfigurable antenna which can change the direction of radiation pattern main lobe

c) Polarization reconfigurable antennas exhibiting a change in polarization from linear-circular and from circular-linear polarization

d) Diversified reconfigurable antenna capable of achieving frequency, pattern, and polarization

Frequency reconfigurable characteristics are achieved by using Electrical/Optical switches which include PIN Diodes, RF MEMS switch, Varactor Diodes, Photo Diodes, Mechanical Actuators, and by using smart materials (Ferrite or Liquid crystals). Also, using these techniques may decrease the efficiency, which may be due to the following reasons:

a) Radiation pattern is affected in the operation of the antenna due to the existence of bias lines

b) Non-linearity and interfering effects of switches

c) Optical switches used in reconfigurable application such as antenna designed on Si-substrate offers losses due to non-linear characteristics and also require additional LASER source for activation

d) It is a very complex task to integrate reconfigurable technique in designing the antenna

10.3   RF PIN Diode and MEMS Switch as Switching Devices

As discussed above, to achieve reconfigurable characteristics either for notched band antennas or for multiband antennas, different switching techniques have been reported. In general, to achieve reconfigurability characteristics RF PIN diodes are used and switching of these diodes can be understood by studying the equivalent circuit model where it offers low impedance in forwarding bias condition, and high impedance in reverse bias condition. Figure 10.1(a) shows the RF PIN diode which can be used as an ON/OFF switch for applications in RF components. When the RF Diode is switched ON, the equivalent circuit can be modeled as a series-connected inductor with a very low resistance of value 4.7Ω. This combination results in very low equivalent impedance and hence operates in ON condition allow the flow of RF signal current. Series resistance designated as Rs can be calculated by

$$R_s = \frac{W^2}{(\mu_n + \mu_p)Q}\Omega \qquad (10.1)$$

Where Q is the charge in coulombs and is given by $Q=I_f\times\tau$ where W is the width of the current region, I_f is forward bias current, τ is carrier lifetime μ_n/μ_p are the mobility of electron and holes. When the PIN diode is in OFF state as shown by Figure 10.1(c), the total equivalent circuit is modeled as a parallel connection of resistance (Rp) and capacitor (CT) which is in turn connected in series with the inductor. This offers very high impedance and the diode is in OFF state blocking RF signal current. Further, is calculated as

$$C_T = \frac{\varepsilon \times A}{W} \qquad (10.2)$$

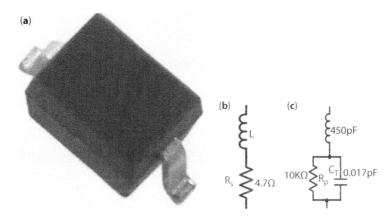

Figure 10.1 (a) Photograph of manufactured PIN diode (Courtesy: NXP Semiconductors) (b) Equivalent circuit model in forwarding bias (c) Equivalent circuit model in reverse bias.

Where ε is the dielectric value and A is the junction diode area.

The above-said conditions of ON/OFF state switching action of the diode are used to reconfigure either notched band antennas or also to control individual resonating band in multiband antenna. Figure 10.2 shows the MEMS switch, which is also one of the optional methods to reconfigure the antenna. RF MEMS SP4T packaged switch inside the slots (ADGM1304) which is also used to reconfigure the antenna is shown in Figure 10.2(a). The shown switch has four ports and all the ports can be used to reconfigure as many as four interfering bands or capable of controlling four resonating bands in a reconfigurable antenna. The internal structure of the SP4T switch demonstrates the presence of a cantilever beam which is fixed at one end and suspended at the other end which is shown in Figure 10.2(b). The upstate and downstate which is controlled by pull-in voltage of a cantilever beam of RF MEMS tune the switch in ON and OFF condition, respectively.

(a)

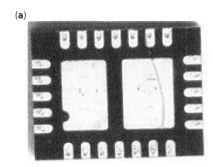

(b)

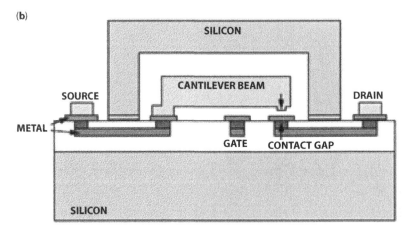

Figure 10.2 (a) Image of packaged SP4T switch (ADGM1304) (b) Cross-sectional view of internal structure of SP4T.

Table 10.1 Comparison of an RF PIN diode and MEMS switch.

Characteristic	PIN diode	MEMS switch
Insertion Loss	<0.8dB	<0.1dB
Power handling capacity	Low	Very low
Isolaion	>45dB	>100dB
Switching speed	10-100ns	µs
Consumption of power	High	Low
Cost	Low	High
Range of Frequency	MHz-GHz	DC-GHz
Excitation/Actuation voltage	3-5V	30-100V

Table 10.1 shows the comparison between RF PIN diode and MEMS switch. It can be seen that from a cost-effective point of view, RF PIN diode overshadows MEMS switch with inheriting qualities such as high switching speed and also a very low requirement of excitation voltage up to 5V.

10.4   Triple Notched Band Reconfigurable Antenna

Figure 10.3 shows the super-wideband monopole antenna which is developed in three stages [37]. Initially, a hexagonal geometry printed on one plane of Rogers RTDuroid substrate and ground on the opposite plane provides working impedance bandwidth of 2.29GHz-13.87GHz and the side of the hexagon patch is calculated by

$$r = \frac{3L_h}{4\pi} \; ; \; L_h = \frac{H}{\sqrt{3}} \tag{10.3}$$

Where r is the equivalent radius of the circle encircling the circumference of the hexagon patch and H is the height of the hexagon patch. Embedding a single ellipse on radiation patch forming Ant. 2 increases working bandwidth and covers 2.85GHz-15.47GHz. The final version of the antenna (Ant. 3) is achieved by adding another ellipse which provides super-wideband bandwidth of 2.85GHz-19.44GHz. After designing the super wideband antenna, it is required for the antenna to mitigate

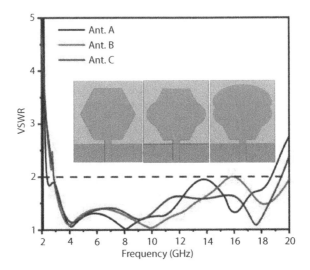

Figure 10.3 Antenna configuration.

interfering bands which include WiMAX, WLAN, and DSS bands. This is achieved by introducing band-notched filters in the antenna. However, different methodologies have been proposed in the literature to achieve wideband antenna with bandstop filters including Complimentary Split Ring Resonator, Electromagnetic Band Gap Structure, parasitic structures, metamaterial structures, using slots and stubs of different shapes like C, T, H, etc. Here, to analyze the above concept, the antenna is shown in Figure 10.3 (Ant. C) is modified by adding a T-type stub placed within an etched rectangular slot encountering WiMAX interference. Also, two C-type slots etched on the radiating patch ensure the removal of WLAN and DSS interfering bands. This study is shown in Figure 10.4(a)-(d) and the respective graph is analyzed in Figure 10.4(e).

As per the observation, the working bandwidth of the antenna with interfering bands is recorded in Table 10.2, which shows comparison results of all the four different working antennas. Hence, it can be concluded, that four different antennas are required to work as SWB, SWB-Single Notch, SWB-Dual Notch, and SWB-Triple notch which makes the PCB very complex in the application environment and also increases the size of the motherboard, which is not desirable as it also increases the cost.

Conversion of the final antenna into reconfigurable characteristics ensures a single antenna maintaining all the characteristics of four antennae in one, which is selected according to the switching condition of PIN diodes. Figure 10.5 shows the triple notched band reconfigurable monopole

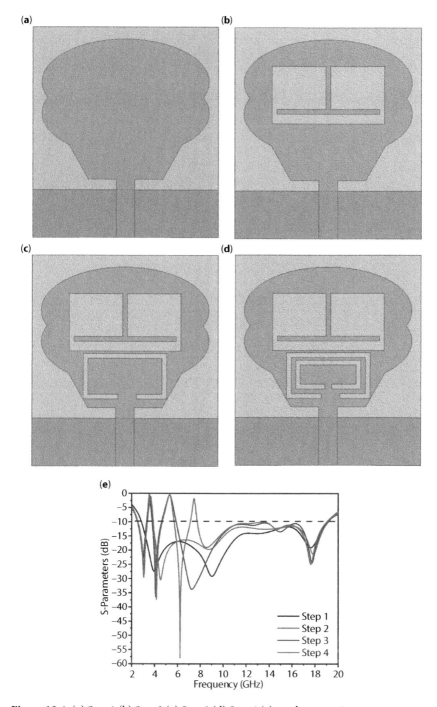

Figure 10.4 (a) Step 1 (b) Step 2 (c) Step 3 (d) Step 4 (e) result comparison.

Table 10.2 Reconfigurable characteristics of reconfigurable notched band super wideband antenna.

PIN diode	ON/OFF			WiMAX (GHz)	WLAN (GHz)	DSS (GHz)	Operating bandwidth (GHz)
	D_1	D_2	D_3				
Mode 1	OFF	ON	ON	✕	✕	✕	✓
Mode 2	ON	ON	ON	✓	✕	✕	✓
Mode 3	OFF	OFF	ON	✕	✓	✕	✓
Mode 4	OFF	ON	OFF	✕	✕	✓	✓
Mode 5	ON	OFF	OFF	✓	✓	✓	✓

WiMAX: Wireless Interoperability for Microwave Access, WLAN: Wireless Local Area Network, DSS: Downlink Satellite System.

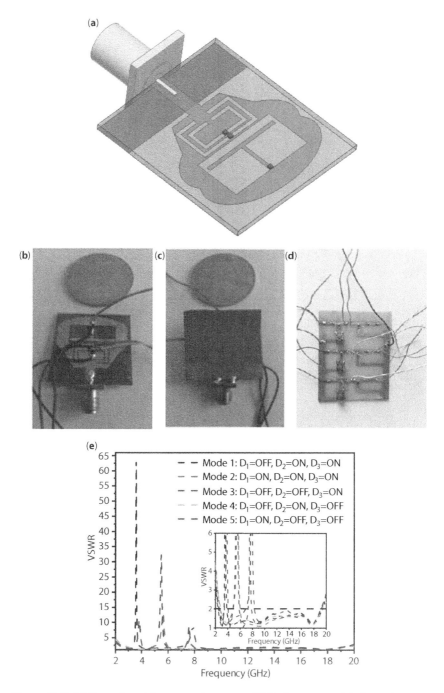

Figure 10.5 (a) Reconfigurable monopole antenna (b) Radiating view (c) Ground view (d) Biasing circuit (e) VSWR results.

Table 10.3 Comparison of notched band reconfigurable monopole antennas.

Ref.	Size (mm²)	Notched bands	Devices/switches used to reconfigure with numbers	Operating bandwidth	Far-field results	
					Gain (dB)	Radiation efficiency (%)
[1]	$0.203\lambda_o \times 0.203\lambda_o$	WLAN	PIN Diode 1	3.06-13.10	2.00-5.23	-
[3]	$0.228\lambda_o \times 0.304\lambda_o$	WLAN DSS	Ideal Switches 2	2.86-12.11	1.95-5.23	-
[5]	$0.203\lambda_o \times 0.203\lambda_o$	WiMAX WLAN DSS	Ideal Switches 4	2.71-12.08	1.89-5.88	-
[6]	$0.120\lambda_o \times 0.180\lambda_o$	WLAN ITS ITU	PIN Diodes 3	3.04-14.52	3.04-5.89	-
[8]	$0.200\lambda_o \times 0.200\lambda_o$	WiMAX WLAN	PIN Diodes 2	3.01-13.62	0.20-0.95	-

(Continued)

Table 10.3 Comparison of notched band reconfigurable monopole antennas. (*Continued*)

Ref.	Size (mm²)	Notched bands	Devices/switches used to reconfigure with numbers	Operating bandwidth	Far-field results	
					Gain (dB)	Radiation efficiency (%)
[10]	$0.280\lambda_o \times 0.280\lambda_o$	WiMAX WLAN ITU	PIN Diodes 3	3.01-12.03	3.01-3.38	-
[11]	$0.302\lambda_o \times 0.358\lambda_o$	WiMAX WLAN DSS	PIN Diodes 3	3.35-39.32	2.21-6.91	72-92%
[12]	$0.220\lambda_o \times 0.280\lambda_o$	WiMAX C Band WLAN	PIN Diodes 3	3.02-14.01	2.04-3.86	80-95%

antenna and band notch characteristics are achieved by using a T-type stub and pair of etched C-shaped slot on the radiating patch as shown by Figure 10.5(a).

The figure also shows the placement of RF pin diodes which are used to achieve reconfigurable characteristics. Figure 10.5(b)-(c) shows soldered PIN diodes and a ground view of the antenna. The external biasing circuit is designed which not only consists of lumped inductors and capacitors but also contains a push button to control the triggering of the PIN diode, as shown in Figure 10.5(d). Reconfigurability of the triple notched band super-wideband antenna can be explained by referring to different modes of operation signified by the ON/OFF state of the diodes used concerning Figure 10.4(e). Mode 1 suggests that when D_1=OFF, D_2=D_3=ON state, all the three filters are deactivated electrically, and hence only super-wideband

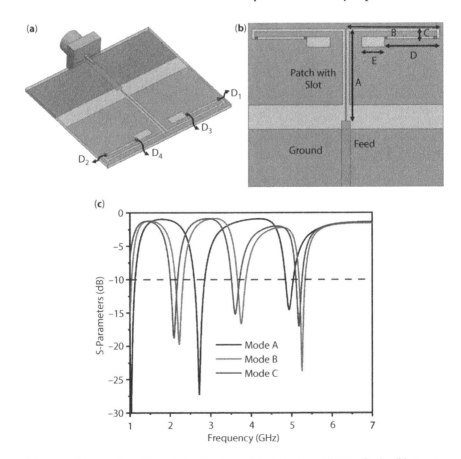

Figure 10.6 Reconfigurable multiband antenna (a) Slant view with PIN diodes (b) Front view with dimensions (c) Reconfigurable modes.

Table 10.4 Multiband reconfigurable antenna.

Modes of PIN diodes	ON/OFF				UMTS (GHz)	WiMAX (GHz)	WLAN (GHz)	Remarks
	D_1	D_2	D_3	D_4				
Mode 1	ON	ON	ON	ON	2.58-2.82	4.84-5.03	-	Two bands
Mode 2	OFF	OFF	ON	ON	2.12-2.30	3.66-3.87	5.18-5.32	Tuning of bands required
Mode 3	OFF	OFF	OFF	OFF	1.99-2.18	3.48-3.69	5.10-5.25	Required bands are obtained

bandwidth is obtained. On the other hand, When D1=ON, D2=D3=OFF, the antenna negotiates all the three interfering bands without comprising the operating super-wideband bandwidth. Similarly, Mode 2-Mode 3 can be explained as per Table 10.2. Table 10.3 shows the comparison of wideband antennas with reconfigurable characteristics. It can be observed that the PIN diode is the general choice for the designer and can easily control notched bands independently depending on the mode of operation.

10.5   Tri-Band Reconfigurable Monopole Antenna

Figure 10.6 shows the configuration of a multiband antenna that resonates for three bands, namely UMTS (Universal Mobile Telecommunication System). WiMAX and WLAN bands [41]. As per the observation from Figure 10.6(a), stepped-microstrip feed is used to match the rectangular patch and the patch is etched with two folded symmetric structures producing three bands of operation. The antenna is designed on an FR4 substrate with a dimension of $0.190\lambda_o \times 0.215\lambda_o$.

Two pairs of diodes (D_1-D_2, D_3-D_4) are placed on the symmetrically folded slot, and their switching action results in the different physical lengths of the slot which affects the impedance bandwidth. The working of the reconfigurable antenna can be analyzed by a discussion of three modes which is tabulated in Table 10.4.

10.6   Conclusions

This chapter discusses the detailed study of reconfigurable antennas which are either used as wideband antennas or multiband antennas. Two reference antennas were considered which have applications in wideband with notched filters and others for multiband applications. Switching mechanisms such as PIN diode, MEMS switch, Varactor diode were discussed and also comparison was made between PIN diode and MEMS switch. It can be concluded that RF PIN diode overshadows MEMS switch due to several advantages; a few of them are high speed of operation, low cost and easy integration on PCB and also uses simple biasing circuit. Two antennas, one super wideband with triple notched band characteristics, and other multiband functional was analyzed to understand the advantages of reconfigurability, and hence maximum power saving can be achieved when these antennas are deployed to fulfill the need of wireless applications.

References

1. S. Ojaroudi, Y. Ojaroudi, and N. Ojaroudi, "Novel design of reconfigurable microstrip slot antenna with switchable band-notched characteristic," *Microwave and Optical Technology Letters*, vol. 57, no. 4, pp. 849-853, 2015.
2. Y. Li and W. Li, "A Circular Slot Antenna with Wide Tunable and Reconfigurable Frequency Rejection Characteristic Using Capacitance Loaded Split-Ring Resonator for UWB Applications," *Wireless Personal Communications*, vol. 78, no. 1, pp. 137-149, 2014.
3. Y. Li, W. Li, and Q. Ye, "A reconfigurable wide slot antenna integrated with sirs for UWB/multiband communication applications," *Microwave and Optical Technology Letters*, vol. 55, no. 1, pp. 52-55, 2013.
4. Y. Li, W. Li, and R. Mittra, "A compact CPW-fed circular slot antenna with reconfigurable dual band-notch characteristics for UWB communication applications," *Microwave and Optical Technology Letters*, vol. 56, no. 2, pp. 465-468, 2014.
5. Y. Li, W. Li, and Q. Ye, "A compact circular slot UWB antenna with multimode reconfigurable band-notched characteristics using resonator and switch techniques," *Microwave and Optical Technology Letters*, vol. 56, no. 3, pp. 570-574, 2014.
6. N. Ojaroudi, N. Ghadimi, Y. Ojaroudi, and S. Ojaroudi, "A novel design of microstrip antenna with reconfigurable band rejection for cognitive radio applications," *Microwave and Optical Technology Letters*, vol. 56, no. 12, pp. 2998-3003, 2014.
7. A. A. Kalteh, G. R. DadashZadeh, M. Naser-Moghadasi, and B. S. Virdee, "Ultra-wideband circular slot antenna with reconfigurable notch band function," *IET Microwaves, Antennas & Propagation*, vol. 6, no. 1, 2012.
8. M. Borhani Kakhki and P. Rezaei, "Reconfigurable microstrip slot antenna with DGS for UWB applications," *International Journal of Microwave and Wireless Technologies*, vol. 9, no. 7, pp. 1517-1522, 2017.
9. T. Singh, K. A. Ali, H. Chaudhary, D. R. Phalswal, V. Gahlaut, and P. k. singh, "Design and Analysis of Reconfigurable Microstrip Antenna for Cognitive Radio Applications," *Wireless Personal Communications*, vol. 98, no. 2, pp. 2163-2185, 2017.
10. E. Nasrabadi and P. Rezaei, "A novel design of reconfigurable monopole antenna with switchable triple band-rejection for UWB applications," *International Journal of Microwave and Wireless Technologies,* vol. 8, no. 8, pp. 1223-1229, 2015.
11. S. Singh, R. Varma, M. Sharma, and S. Hussain, "Superwideband Monopole Reconfigurable Antenna with Triple Notched Band Characteristics for Numerous Applications in Wireless System," *Wireless Personal Communications*, vol. 106, no. 3, pp. 987-999, 2019.
12. M. I. Magray, K. Muzaffar, Z. Wani, R. K. Singh, G. S. Karthikeya, and S. K. Koul, "Compact frequency reconfigurable triple band notched monopole

antenna for ultrawideband applications," *International Journal of RF and Microwave Computer-Aided Engineering*, vol. 29, no. 11, 2019.

13. G. Jin, L. Li, W. Wang, and S. Liao, "A wideband polarization reconfigurable antenna based on optical switches and C-shaped radiator," *Microwave and Optical Technology Letters*, vol. 62, no. 6, pp. 2415-2422, 2020.

14. M. M. Shirkolaei and M. Jafari, "A new class of wideband microstrip falcate patch antennas with reconfigurable capability at circular-polarization," *Microwave and Optical Technology Letters*, vol. 62, no. 12, pp. 3922-3927, 2020.

15. K. R. Shashikant and A. Kulkarni, "Reconfigurable Patch Antenna Design Using Pin Diodes and Raspberry PI for Portable Device Application," *Wireless Personal Communications*, vol. 112, no. 3, pp. 1809-1828, 2020.

16. I. Yeom, J. Choi, S.-s. Kwoun, B. Lee, and C. Jung, "Analysis of RF Front-End Performance of Reconfigurable Antennas with RF Switches in the Far Field," *International Journal of Antennas and Propagation*, vol. 2014, pp. 1-14, 2014.

17. A. K. Gangwar and M. S. Alam, "Frequency reconfigurable dual-band filtenna," *AEU - International Journal of Electronics and Communications*, vol. 124, 2020.

18. S. Genovesi, A. Monorchio, M. B. Borgese, S. Pisu, and F. M. Valeri, "Frequency-Reconfigurable Microstrip Antenna with Biasing Network Driven by a PIC Microcontroller," *IEEE Antennas and Wireless Propagation Letters*, vol. 11, pp. 156-159, 2012.

19. M. Ameen, O. Ahmad, and R. K. Chaudhary, "Single split-ring resonator loaded self-decoupled dual-polarized MIMO antenna for mid-band 5G and C-band applications," *AEU - International Journal of Electronics and Communications*, vol. 124, 2020.

20. H. Koc Polat, M. D. Geyikoglu, and B. Cavusoglu, "Modeling and validation of a new reconfigurable patch antenna through equivalent lumped circuit-based design for minimum tuning effort," *Microwave and Optical Technology Letters*, vol. 62, no. 6, pp. 2335-2345, 2020.

21. T. A. Elwi, "Remotely controlled reconfigurable antenna for modern 5G networks applications," *Microwave and Optical Technology Letters*, 2020.

22. A. Boufrioua, "Frequency Reconfigurable Antenna Designs Using PIN Diode for Wireless Communication Applications," *Wireless Personal Communications*, vol. 110, no. 4, pp. 1879-1885, 2019.

23. W. A. Awan *et al.*, "A miniaturized wideband and multi-band on-demand reconfigurable antenna for compact and portable devices," *AEU - International Journal of Electronics and Communications*, vol. 122, 2020.

24. T. Ali, M. Muzammil Khaleeq, and R. C. Biradar, "A multiband reconfigurable slot antenna for wireless applications," *AEU - International Journal of Electronics and Communications*, vol. 84, pp. 273-280, 2018.

25. M. Kanagasabai *et al.*, "On the design of frequency reconfigurable tri-band miniaturized antenna for WBAN applications," *AEU - International Journal of Electronics and Communications*, vol. 127, 2020.

26. A. A. Palsokar and S. L. Lahudkar, "Frequency and pattern reconfigurable rectangular patch antenna using single PIN diode," *AEU - International Journal of Electronics and Communications*, vol. 125, 2020.

27. D. R. Sandeep, N. Prabakaran, B. T. P. Madhav, K. L. Narayana, and Y. P. Reddy, "Semicircular shape hybrid reconfigurable antenna on Jute textile for ISM Wi-Fi , Wi-MAX , and W-LAN applications," *International Journal of RF and Microwave Computer-Aided Engineering*, vol. 30, no. 11, 2020.

28. Y. Nafde and R. Pande, "Design and Analysis of Resistive Series RF MEMS Switches Based Fractal U-Slot Reconfigurable Antenna," *Wireless Personal Communications*, vol. 97, no. 2, pp. 2871-2886, 2017.

29. F. Zadehparizi and S. Jam, "Frequency Reconfigurable Antennas Design for Cognitive Radio Applications with Different Number of Sub-bands Based on Genetic Algorithm," *Wireless Personal Communications*, vol. 98, no. 4, pp. 3431-3441, 2017.

30. S. Tripathi, A. Mohan, and S. Yadav, "A compact frequency-reconfigurable fractal UWB antenna using reconfigurable ground plane," *Microwave and Optical Technology Letters*, vol. 59, no. 8, pp. 1800-1808, 2017.

31. I.H. Idris, M.R. Hamid, M.H. Jamaluddin, M.K.A. Rahim, J.R. Kelly and H.A. Majid, "Single-, Dual- and Triple-band Frequency Reconfigurable Antenna," *Radioengineering*, vol. 23, no. 3, pp. 805-811, 2014.

32. A. Asghari, N. Azadi-Tinat, H. Oraizi, and J. Ghalibafan, "Wideband Frequency-Reconfigurable Antenna for Airborne Applications," *Wireless Personal Communications*, vol. 109, no. 3, pp. 1529-1540, 2019.

33. R. K. Saraswat and M. Kumar, "Implementation of Vertex-Fed Multiband Antenna for Wireless Applications with Frequency Band Reconfigurability Characteristics," *Wireless Personal Communications*, 2020.

34. M. Jenath Sathikbasha and V. Nagarajan, "Design of Multiband Frequency Reconfigurable Antenna with Defected Ground Structure for Wireless Applications," *Wireless Personal Communications*, vol. 113, no. 2, pp. 867-892, 2020.

35. V. Arun and L. R. Karl Marx, "Micro-controlled Tree Shaped Reconfigurable Patch Antenna with RF-Energy Harvesting," *Wireless Personal Communications*, vol. 94, no. 4, pp. 2769-2781, 2017.

36. N. P. Gupta and M. Kumar, "Development of a Reconfigurable and Miniaturized CPW Antenna for Selective and Wideband Communication," *Wireless Personal Communications*, vol. 95, no. 3, pp. 2599-2608, 2017.

37. M. Sharma, Y.K. Awasthi and H. Singh, "Compact multiband planar monopole antenna for Bluetooth, LTE, and Reconfigurable UWB applications including X-Band and Ku-band Applications," *International Journal of RF and Microwave Computer-Aided Engineering*, April 12, 2019.

38. Ojaroudi Parchin, H. Jahanbakhsh Basherlou, Y. I. A. Al-Yasir, A. M. Abdulkhaleq, and R. A. Abd-Alhameed, "Reconfigurable Antennas: Switching Techniques—A Survey," *Electronics*, vol. 9, no. 2, 2020.

39. N. Kumar, P. Kumar, and M. Sharma, "Reconfigurable Antenna and Performance Optimization Approach," *Wireless Personal Communications*, vol. 112, no. 4, pp. 2187-2212, 2020.

40. K. Pedram, J. Nourinia, C. Ghobadi and M. Karamirad, "A multiband circularly polarized antenna with simple structure for wireless communication system," *Microwave Optical Technology Letters*, vol. 59, pp. 2290-2297, 2017.

41. A.R. Jalali, J. Ahamdi-Shokouh and S.R. Emadian, "Compact Multiband Monopole Antenna for UMTS, WiMAX and WLAN Appications," *Microwave Optical Technology Letters*, vol. 15, no. 4, pp. 844-847, 2016.

42. N. Ojaroudi Parchin, H. Jahanbakhsh Basherlou, Y. Al-Yasir, R. Abd-Alhameed, A. Abdulkhaleq, and J. Noras, "Recent Developments of Reconfigurable Antennas for Current and Future Wireless Communication Systems," *Electronics*, vol. 8, no. 2, 2019.

43. G. K. Pandey, H. S. Singh, P. K. Bharti, and M. K. Meshram, "Design and analysis of multiband notched pitcher-shaped UWB antenna," *International Journal of RF and Microwave Computer-Aided Engineering*, vol. 25, no. 9, pp. 795-806, 2015.

11

IoT World Communication through Antenna Propagation with Emerging Design Analysis Features

E.B. Priyanka[1]* and S. Thangavel[2]

[1]Department of Automobile Engineering, Kongu Engineering College, Erode, India
[2]Department of Mechatronics Engineering, Kongu Engineering College, Erode, India

Abstract

Remote innovation is significantly affected by the high-level mechanical improvement of the reception apparatus. With the quick development of the Internet of Things (IoT) application in the cutting-edge correspondence framework, interest in miniature receiving wires is expanding. Microstrip fix radio wires are commonly utilized in IoT applications because of their similarity. The critical advantage of the microstrip reception apparatus is that it permits simple joining into IoT gadgets. IoT conditions are persistently evolving. Changes may come from the administration, availability, or on the other hand actual layers of the IoT engineering. Subsequently, to work suitably, the framework needs to progressively adjust to its current circumstance. This section addresses the difficulties of dealing with the consideration of new antenna designs and antenna transient association, by methods for dynamic transformations consolidated into our proposed programming engineering for versatile IoT frameworks. To oversee dynamic transformations, we expand the reference IoT engineering with our particular segments. In particular, various points of view to consider are ontologies and occurrences to the antenna to the space information; a coordinating calculation to pair administrations and IoT gadgets, considering their practical necessities, quality ascribes and sensors properties; and a match update calculation utilized at whatever point sensors become (un)available.

Keywords: Antenna, IoT, data propagation, antenna designs

Corresponding author: priyankabhaskaran1993@gmail.com

Prashant Ranjan, Dharmendra Kumar Jhariya, Manoj Gupta, Krishna Kumar, and Pradeep Kumar (eds.)
Next-Generation Antennas: Advances and Challenges, (185–202) © 2021 Scrivener Publishing LLC

11.1   Introduction

As with upgraded technology, the web of things towards wireless has shown quick development, and it has changed circumstances more drastically than at any time in recent memory. The IoT depends on a worldview in which all gadgets can be part of a web network of interconnected routers [1]. It implies that we can without much of a stretch get some information started from the end gadgets, and produce more important data from a post-handling system dependent on the immense measure of information. As of late, IoT administrations appear in our day-by-day lives in assorted areas, for example, smart home, home wellbeing, brilliant industrial facility, and smart city [2]. These administrations are essentially zeroed in on the best way to improve accommodation and assure more wellbeing for all. To accomplish the idea of a keen IoT world, the deliberate assembly should be done before by advancement in rudimentary innovation, including remote correspondences, stage, information base, information preparing calculations, and so on. Among these exploration territories, we essentially center around how to execute remote correspondences framework [3].

Considering that the framework highlight generally decides framework costs, it is fundamental to painstakingly pick what kind of remote correspondence innovation is to be used. The framework inclusion decides the number of sensor hubs every entryway can interface with. A sensor hub synthesis for sensor organization comprises different sensors, perused out incorporated circuits and antenna supporting handsets. In the previous few years, an assortment of radio innovation has appeared for low force-wide zone [4]. Contrasted with regular remote correspondences, it ensures an extensively long battery lifetime and wide inclusion. It results that nearly all the on-the-ground physical antenna is firmly associated, and it adds to quick development in IoT administrations advancement. The sensor organization can be grouped into two classes compared to the area of information asset: an IoT framework towards the upgraded data transmission [5]. It implies that the IoT framework depends on underground sensors to assemble information from all the underneath substrate things, for example, mugginess, heat energy, etc. As per the framework classification, the framework creators ought to painstakingly decide how to circulate the structure blocks actually [6].

Terahertz recurrence band as being relevant for the condition of craftsmanship correspondences because the necessity for higher information rates whenever anyplace is to be expanded sooner rather than later with an expansion in the quantity of portable associated antenna design for data transmission with bolt transmitter is shown in Figure 11.1 [6].

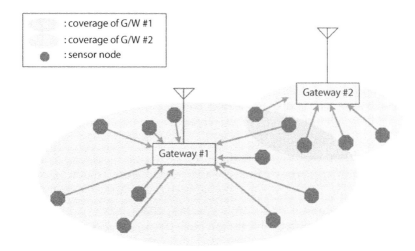

Figure 11.1 Sensor node connected with antenna coverage area with the gateway.

By analyzing the recent trends in millimeter-wave innovation is arriving at business arrangements and still spurred by the deficient data transmission, hence the terahertz (THz) spectrum is imagined as the approaching wilderness for correspondence. Remote correspondence is requested better channel limit with a high information rate in the cutting-edge time. To satisfy the undertaken requests, the MIMO-correspondence frameworks with THz range are needed for enhanced information speed in (Tbps) [7]. Towards the predominant of increased throughput per gadget (from various Gbps to a few Tera-bps) together with per region productivity (bps/km²). It is additionally anticipated that the world month-to-month traffic in cell phones is towards the next revolution of wireless bandwidth.

Thus, the subsequent segment interconnects administrations and gadgets. The interchange between the two parts can be utilized to adjust the association between gadgets (e.g., in the event of disappointments). At last, beat the board, utilized in conveyed frameworks, includes the checking and goal of hubs to oversee hubs joining and leaving the organization. The goal procedures utilized fluctuate in their expectation to fix the organization at whatever point such occasions happen [8]. Prescient procedures use hub data to figure out which hubs to use in the organization or whether they should be supplanted when they come up short. Three fundamental techniques are utilized in the writing to supplant bombing hubs. The arbitrary substitution methodology replaces a bombed hub with some other hub accessible in the framework. The prescient substitution technique is utilized to supplant bombed hubs with a hub from a predefined list (as a rule considering hubs' uptime). Also, the inclination list technique positions hubs as per an inclination request, and uses a subset of such hubs to impact substitutions. Revelation and relationship among gadgets and administrations utilize a metaphysics case coordinating calculation as the initial step to give variation to transient gadgets [9]. Administrations' necessities are coordinated with IoT gadgets' utilitarian and non-useful properties, assembling a set of the most appropriate gadgets, which is endured on the execution guidepost's vault. In any case, after fundamental experimentation, we disposed of the utilization of semantic coordinating because it didn't generously improve the coordinating exactness, while it harmed reaction time. Utilizing the excess coordinating procedures, we think about the useful and non-useful properties between IoT gadgets and web administrations to locate the most fitting match that satisfies the administration's necessities among every accessible gadget [10].

The structure of the chapter is as follows: Section 11.2 describes the design and analysis of the MIMO antenna, Section 11.3 enumerates the measurement consideration and projection of the 3D pattern for the IoT module, and Section 11.4 shows the comparison of antenna various aspects of symmetry towards IoT data communication surveyed with conclusions.

11.2   Design and Parameter Analysis of Multi-Input Multi-Output Antennas

Variety execution of the planned MIMO receiving wire is decided regarding significant boundaries to judge how many measures of electric field relocated from the energized reception apparatus to the remainder of the receiving wire in the MIMO framework [11]. It is determined through S-boundaries utilizing condition and this methodology is substantial when a radio wire is lossless and supremacy is consistently conveyed through the reception apparatus length, i.e., for high productivity antennas [12]. Upon the uncorrelated multi-input multi-output receiving wire framework, the estimation of ECC should be zero yet their pragmatic incentive for a good activity in the MIMO correspondence framework is ≤ 0.23. The mathematical estimation of ECC of planned microstrip antenna receiving wire in SWB data transmission is noticed under 0.0074 as shown in Figure 11.2 [13].

Subsequently, the above outcomes show an extremely low relationship between the two contiguous reception apparatuses, which shows a great variety of execution in the proposed L-formed decoupling structure planning. The channel limit is a significant boundary to pass judgment on the information pace of the MIMO framework [14]. MIMO framework is

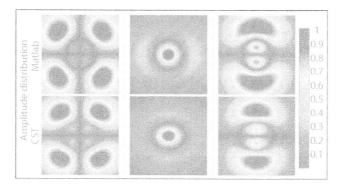

Figure 11.2 Amplitude distribution of microstrip antenna with two analyzing platforms with sample datasets.

utilized to improve the channel limit and the exhibition of transfer speed and SNR (sign to clamor) proportion contrast and the SISO framework. The channel limit of the planned 2-reception apparatus component is calculated. The principal target of the MIMO receiving wire framework is to improve the channel limit according to the prerequisite. The channel limit of the MIMO framework is dependent upon various reception apparatus components associated with the MIMO framework and the measure of the relationship between reception apparatus components [15]. If we increment the quantity of receiving wire components, at that point channel limit increments directly. Yet, the relationship factor and number of receiving wire components in the MIMO framework is built then the channel limit misfortunes (CCL) are likewise expanding. Subsequently, the variety execution of proposed reception apparatus self-important and channel limit misfortunes are calculated. Where mean the character lattice and several communicating and getting reception apparatuses separately, SNR is the proportion of signs feed at reception apparatus port to the channels clamor [16]. For Rayleigh blurring climate for example SNR =36 dB. The quantity of communicating reception apparatus components is meant by k. H and H* are the channel lattice and its Hermitian render of a matrix. In the 2-port reception apparatus framework, neighboring radio wire components interfere with one another when they are working simultaneously. This interference impacts the generally alluring increase, effectiveness, and data transmission of the proposed reception apparatus [17]. Consequently, the genuine MIMO receiving wire framework execution won't be anticipated by S-boundaries just, so another significant boundary (for example TARC) of the MIMO radio wire framework is investigated to approve the variety execution. TARC is characterized as "the square base of the proportion of all-out reflected capacity to the absolute occurrence power" [18].

11.3   Measurement Analysis in 3D Pattern with IoT Module

To execute a correlated wireless framework, prior consideration must be given to various rates of reception apparatus towards the pointed motivation. As shown in Figure 11.3, the ordinary middleware IoT frameworks can be partitioned to implant its reception apparatus for remote interchanges. Initially, receiving wire is exclusively introduced on routers with the gateway to send towards sensor nodes [19]. At that point, the radio wire displays its exhibitions, including pick up, effectiveness, radiation design, paying little heed to underground climate, which is isolated from different squares of underground sensor hub. The most straightforward approach to actualize a middleware IoT framework; in any case, there exist wellbeing issues concerning structural designing perspective. Taking into account that the receiving wire also, its bundle distends around the bandwidth

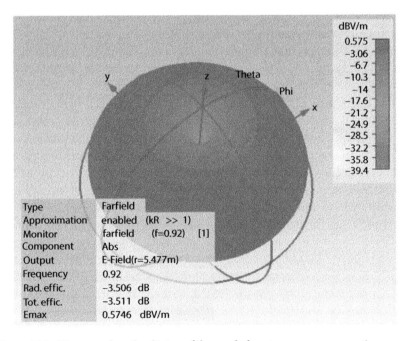

Figure 11.3 3D pattern-based radiation of the manhole antenna wave propagation.

allocations, it is hard to ensure the street wellbeing with isolated sensor hub frameworks [20]. Besides, the receiving wire of every sensor hub can be covered or set underground. The depth of signal to noise ratio with the gateway to the sensor node by antenna ranging will lead to the data transmission in a faster manner due to the involvement of array antenna design supporting base stations [21].

Likewise, the sewer vent concealment frame, ended with cast iron combined graphite materials, blocks electromagnetic waves emanated from the reception apparatus, and in this manner seriously debases the receiving wire exhibitions [22]. Expecting that the sensor hubs are embedded with indistinguishable scattering, the sensor organization can be diversely designed as per its radio wire inclusion region. As the radio wires on the left side give more extensive inclusion territory contrasted with those on the correct side, the single door is sufficient to help all the sensor hubs [23, 24]. For the enormous sensor hubs in metropolitan climate, the framework building cost incredibly increments while abusing the covered sort reception apparatus. It is essential to think about the two qualities of a street wellbeing and inclusion region. It implies that two kinds of regular receiving wires are unseemly for developing savvy and reasonable framework [25]. A recently proposed radio wire for remote sensor organization: a sewer vent cover with coordinated reception apparatus. As advancing into the implanted structure, the receiving wire inalienably guarantees the security of vehicles and people disregarding the sewer vent cover [26].

As the sewer vent cover formed reception antenna designs with similar tallness around itself, it can't maintain a strategic distance from that proliferation invalid emerges along with the ground surface level. The connection between receiving wire height point and inclusion region. With a rising point of θ1 and θ2 from the beginning level, the ground-level reception apparatus has a diverse inclusion region of every sensor hub [27]. It suggests that the lower height point gives the more extensive inclusion region and high dependability for correspondence connect. it is imperative to think about the two attributes of a street wellbeing and inclusion zone. It implies that two sorts of regular reception apparatuses are wrong for developing a reasonable framework. As advancing into the installed structure, the reception apparatus innately guarantees the security of vehicles and people disregarding the sewer vent cover [28]. As the sewer vent cover molded reception apparatus designs with a similar stature around itself, it can't stay away from that proliferation invalid emerges along with the ground surface level as shown in Figure 11.4 with 2D pattern [29].

With a rising point of θ1 and θ2 starting from the earliest stage level, the ground-level reception apparatus has a distinctive inclusion territory

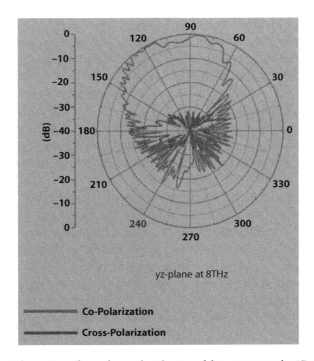

Figure 11.4 Polarization effects of wave distribution of the antenna in the 2D pattern.

of every sensor hub. It infers that the lower rise point allows the more extensive inclusion region and high dependability for correspondence link [30]. Moreover, the calculation channels all correspondences under a given edge. The remaining correspondences are recorded in the execution of the guidepost. The first step, called Functional coordinating (syntactic comparability) in the figure, executes syntactic procedures to break down the similitude of three gadgets' properties as follows: (1) classification (i.e., the portrayal of the actual variable estimated by the gadget); (2) unit of the variable; and (3) gadget area [31]. The second and third steps, alluded to as System properties and QoS properties in the figure, examine the similitude of non-utilitarian properties that depict either the physical (i.e., gadgets) or the virtual (i.e., administrations) world [32]. This means work the similitude results with the pre-designed weight boundaries. The setup is picked in the approaching solicitation. The two stages execute type-based and esteem-based coordinating. The Instance coordinating encourages us to discover elective associations among administrations and IoT gadgets dependent on the best matches of utilitarian and non-useful properties [33]. This segment offers a calculation to oversee transient gadgets and maintain a strategic distance from administration interruptions because of foundation updates and glitches. Given a gadget bound to assistance, any remaining identical gadgets in the climate (i.e., gadgets fulfilling a similar useful and non-utilitarian prerequisites) are kept in our design so that the progress from an imperfect or breaking down gadget (Dold) to another gadget (Drew) giving the data is smooth [34]. On occasion, in the SUDS model, if the stream sensor falls flat (e.g., due to climate harm), another gadget in the climate can be utilized to give the necessary degree of data. Gadgets' portrayal depends on philosophical occurrences; accordingly, the self-sufficient association among gadgets and administrations is set up dependent on the consequence of a cosmology example coordinating calculation. There is a wide range of cosmology coordinating procedures, for example, semantic, syntactic, scientific classification based, model-based, and diagram-based heuristics [35].

11.4   Comparison of Antenna Design Concerning the IoT Data Transmission

The IoT standard was made to empower PCs to speak with their current circumstance and to get information about antennas and objects with no human communication. The reception apparatus is thought of as one of the main parts of a remote correspondence framework since it assists with improving the general execution, furthermore, a prerequisite of the framework when it is all around planned [36]. Microstrip fix reception apparatuses are for the most part utilized in the IoT applications due to their similarity, low profile, lightweight and exactness applications in the microwave recurrence locale. The upside of receiving wire scaling down is that it permits simple reconciliation into routinely little and smaller IoT antenna. The pentagonal fix is the best in light of its exhibition contrasted with another fix. Be that as it may, it was expressed that rectangular fix radio wire is most practical, has tantamount exhibitions and simpler to fabricate as microstrip receiving wires [37]. The increase and data transmission of a straightforward microstrip receiving wire is known to be 1.8 dB what's more, under 1.3% separately. In any case, the appropriateness for IoT applications was not decided. There is no particular transmission capacity necessity in the IoT application, as long the recurrence of the scope of the receiving wire is inside 4.99 GHz. To assess these cases, this work was finished. Accordingly, a thorough report was directed to examine the presentation of microstrip reception apparatus for IoT application [38]. The examination was made between two kinds of microstrip radio wire, which was the customary microstrip reception apparatus and enhanced microstrip receiving wire with a U-formed structure. The exhibitions of these two receiving wires were assessed as far as transfer speed, pick up and return of misfortune. Figure 11.5 describes the radiation pattern for a case undertaken on a sample analysis of an antenna with high gain [39].

Stacking space on receiving wire fix assists with decreasing receiving wire size and assists with improving data transfer capacity and gain. These stacked spaces in the microstrip fix bring about contrasts in the current appropriation as they cut along the non-transmitting edge, consequently expanding the receiving wire's opposition and limit [40]. This expansion happens because the current way length has been expanded because of the expansion of spaces to the receiving wire fix, driving to the option of extra capacitance and obstruction in the circuit. What's more, these additional openings make the surface flows wander, accordingly expanding the electrical length of the receiving wire without modifying its all-out measurements [41]. These methods are regularly sensitive to recurrence. For this work, the CST programming 2014 form is utilized for planning also, reproduction. For this work, rectangular fix microstrip radio wire, full at a recurrence of 6.1 GHz, with a substrate measurement of 23.9 mm*19.2 mm and fix measurement of 14.3 mm*10.7 mm was planned [42].

The substrate which was utilized for the plan of the radio wire is FR-4 which has a dielectric consistent of $Er = 3.97$, and a substrate thickness of 2.21 mm and a misfortune digression of 0.032. The recipe utilized in

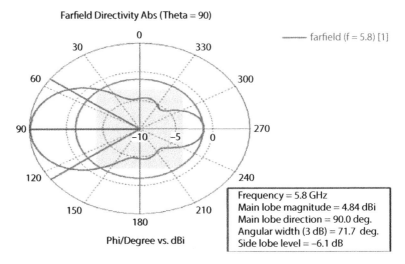

Figure 11.5 Parameter estimation in far-field directivity with gain estimation.

computing the measurement for this regular receiving wire was adjusted by the conditions [43]. The conditions that were alluded are fixed width condition (W), successful dielectric steady condition ($\mathcal{E}$), viable length condition (Leff), standardized expansion long (L), and feed length equation. This increment happens because the current way length has been expanded because of the expansion of openings to the receiving wire fix, driving to the option of extra capacitance and opposition in the circuit [44]. Also, these additional spaces make the surface flows wander, consequently expanding the electrical length of the reception apparatus without adjusting its all-out dimensions as shown in Figure 11.6.

In case 2, the advanced receiving wire displayed an improved working recurrence of 4.978 GHz, which appeared an improvement of 0.0119 GHz. As far as return misfortune, for case 1, the regular reception apparatus displayed a return deficiency of 19.97dB. On the other hand, for case 2, the enhanced radio wire showed an improved return deficiency of 23.78 dB [45]. Subsequently, it is found that the energy utilization for the streamlined receiving wire is in this manner more compelling contrasted with the standard receiving wire, as less force is returned by the streamlined receiving wire. Regarding pick up, for case 1, the regular receiving wire displayed pick up the estimation of 2.7 dB [46–49]. Thus, the advanced reception apparatus can be arranged as an amazing sign radio wire that can impart or get an incredible sign in a provided guidance inferable from its

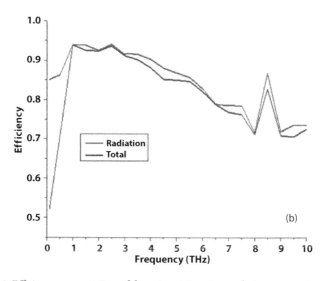

Figure 11.6 Efficiency computation of the microstrip antenna design concerning frequency.

more noteworthy addition esteem. Furthermore, as far as radiation design, the upgraded radio wire demonstrated an improved Yagi radiation design with principal projection size estimation of 5.467 dBi [50–54], primary flap heading of 49.7 degrees, precise width of 112.3 degrees and side projection level of 5.98dB. This is contrasted with the ordinary antenna which showed primary flap greatness estimation of 3.987 dBi, a fundamental projection heading of 87.0-degree, precise width of 69.92 degrees and a side projection level of 5.76 dB [55]. Moreover, the outcomes acquired from the recreation of the receiving wire fulfill the WLAN norms.

11.5   Conclusions

In the near future, remote correspondence items and electronic contraptions will become essential for human existence. Correspondence frameworks need reception apparatuses that work with a multiband and wideband with required boundaries like polarization and gain. The fundamental purpose of this work is to deliver high bar shaping with the guide of common coupling among the four antenna components to support upgraded travel limit and enable the correspondence transmission capacities at exceptionally enormous information rates for 5G Technology. In the perspective on moderating the multipath blurring with previously mentioned standards. The planned radio wire is built up as a MIMO fix reception apparatus with wide attributes. It works the recurrence space from 1.98 GHz to 5.26 GHz. The reception apparatus is created with FR4 material with a dielectric steady of 4.4, misfortune digression of 0.018 and a thickness of 1.45 mm. The suggested configuration has 4 monopole reception antennas. Each monopole receiving antenna has a round fix with a range of 4.8 mm to evading impedance. A super-wideband 2-components MIMO receiving antenna has been intended for THz applications. This receiving wire has accomplished an exceptionally high pinnacle gain of 23dBi including stable order radiation designs with worthy radio wire and variety execution. The exhibitions of the MIMO reception apparatus have demonstrated that it is a decent model for THz remote correspondence and examining networks. Performance properties assume a vital part in meeting applications' destinations. Regardless, the incorporation of such properties doesn't raise the specialized multifaceted nature of characterizing an IoT framework, while it makes unequivocal attributes characteristic for the difficult space. Given the choice, by using weight change on the coordinating calculation to appoint needs on framework properties as per the application goals.

Acknowledgement

The author would like to acknowledge the CSIR-HRDG EMR section for partially supporting the research work by Senior Research Fellowship with grant no.08/678(0001)2k18 EMR.

References

1. Priyanka, E. B., S. Thangavel, D. Venkatesa Prabu. Fundamentals of Wireless Sensor Networks Using Machine Learning Approaches: Advancement in Big Data Analysis Using Hadoop for Oil Pipeline System with Scheduling Algorithm. *Deep Learning Strategies for Security Enhancement in Wireless Sensor Networks*. IGI Global, 2020. 233-254. DOI: 10.4018/978-1-7998-5068-7.ch012

2. Priyanka, E. B., Thangavel, S., Pratheep, V. G. (2020). Enhanced Digital Synthesized Phase Locked Loop with High Frequency Compensation and Clock Generation. *Sensing and Imaging*, 21(1), 1-12. https://doi.org/10.1007/s11220-020-00308-0

3. Pazin, Leviatan Y. Reconfigurable slot antenna for switchable multiband operation in a wide frequency range. *IEEE Antennas Wireless Propag Lett* 2013;12(Feb):329–33.

4. Pazin L, Leviatan Y. Reconfigurable rotated-t slot antenna for cognitive radio systems. *IEEE Trans Antennas Propag* 2014;62(5):2382–7.

5. Aboufoul T, Parini C, Chen X, Alomainy A. Pattern-reconfigurable planar circular ultra-wideband monopole antenna. *IEEE Trans Antennas Propag* 2013;61(10):4973–80.

6. Kumar Rajeev, Vijay Ritu. Quadrilateral patch and slot based optimal frequency agile antenna for cognitive radio system. *Int J RF Microwave Comput Aided Eng* Feb 2018;28(2).

7. Khidre A, Yang F, Elsherbeni AZ. A patch antenna with a varactor-loaded slot for reconfigurable dual-band operation. *IEEE Trans Antennas Propag* 2015;63 (2):755–60.

8. Mansoul A, Ghanem F, Hamid MR, Trabelsi M. A selective frequency reconfigurable antenna for cognitive radio applications. *IEEE Antennas Wirel Propag Lett* 2014; 13:515–8.

9. Hamid MR, Gardner P, Hall PS, Ghanem F. Switched-band vivaldi antenna. *IEEE Trans Antennas Propag* 2011;59(5):1472–80.

10. Punjala SS, Pissinou N, Makki K. A multiple resonant frequency circular reconfigurable antenna investigated with wireless powering in a concrete block. *Int J Antennas Propag* 2015; 2015(April): 9 pages, 413642.

11. Kumar R, Vijay R. A frequency agile semicircular slot antenna for cognitive radio systems. *Int J Microwave Sci Technol* 2016;2016(April):11pages 2648248.

12. Majid HA, Rahim MKA. Frequency-reconfigurable microstrip patch-slot antenna. *IEEE Antennas Wirel Propag Lett* 2013;12(Feb):218–20.

13. Majid HA, Rahim MKA. A compact frequency-reconfigurable narrowband microstrip slot antenna. *IEEE Antennas Wirel Propag Lett* 2012;11(June):616–9.

14. Priyanka EB, Maheswari C, Thangavel S. Proactive Decision Making Based IoT Framework for an Oil Pipeline Transportation System. In *International conference on Computer Networks, Big data and IoT 2018 Dec 19* (pp. 108-119). Springer, Cham. https://doi.org/10.1007/978-3-030-24643-3_12

15. Priyanka, E.B., Maheswari, C. and Thangavel, S., 2018. IoT based field parameters monitoring and control in press shop assembly. *Internet of Things*, 3, pp. 1-11. https://doi.org/10.1016/j.iot.2018.09.004

16. Priyanka, E. B., Maheswari, C., Thangavel, S. & Bala, M. P. (2020) Integrating IoT with LQR-PID controller for online surveillance and control of flow and pressure in fluid transportation system. *Journal of Industrial Information Integration*, 17, 100127. https://doi.org/10.1016/j.jii.2020.100127

17. Subramaniam, T., Bhaskaran, P. (2019). Local intelligence for remote surveillance and control of flow in fluid transportation system. *Advances in Modelling and Analysis C*, Vol. 74, No. 1, pp. 15-21. https://doi.org/10.18280/ama_c.740102.

18. Selvam YP *et al*. A low-profile frequency- and pattern-reconfigurable antenna. *IEEE Antennas Wirel Propag Lett* 2017; 16:3047–50.

19. Romputtal A, Phongcharoenpanich C. Frequency reconfigurable multiband antenna with embedded biasing network. *IET Microwaves Antennas Propag* 2017;11(10):1369–78.

20. C. Maheswari, E.B. Priyanka, S. Thangavel, P. Parameswari. Development of unmanned guided vehicle for material handling automation for industry 4.0, in *International Journal of Recent Technology & Engineering*, Volume 7, Issue 4s, November 2018, pp. 428-432.

21. E.B. Priyanka, S. Thangavel, P. Parameswari, Automated Pay and Use Browsing and Printing Machine in *International Journal of Innovative Technology and Exploring Engineering (IJITEE)*, Volume 8, Issue 11S, September 2019, pp. 148-152.

22. V.G. Pratheep, E.B. Priyanka, R. Raja. Design and fabrication of 3-axis welding robot in *International Journal of Innovative Technology and Exploring Engineering (IJITEE)*, Volume 8, Issue 11, September 2019, pp. 1588-1592.

23. E.B. Priyanka, S. Thangavel, P. Parameswari, Collision Waring System Using RFID in Automotives in *International Journal of Innovative Technology and Exploring Engineering (IJITEE)*, Volume 8, Issue 11S, September 2019, pp. 153-158.

24. Varamini G, Keshtkar A, Naser-Moghadasi M. Compact and miniaturized microstrip antenna based on Fractal and metamaterial loads with

reconfigurable qualification. *Int J Electron Commun* 2017. https://doi.org/10.1016/j.aeue.2017.08.057.

25. Anantha B, Merugu L, Rao S. A quad-polarization and frequency reconfigurable square ring slot loaded microstrip patch antenna for WLAN applications. *Int J Electron Commun* 2017; 78:15–23.

26. Bharathi A, Lakshminarayana M, Somasekhar Rao PVD. A novel single feed frequency and polarization reconfigurable microstrip patch antenna. *Int J Electron Commun (AEU)*, Elsevier. 2017; 72:8–16.

27. Ali T, Khaleeq MM, Biradar RC. A multiband reconfigurable slot antenna for wireless application. *Int J Electron Commun (AEU)* 2018; 84:273–80.

28. Bhaskaran, P. E., Chennippan, M., & Subramaniam, T. (2020). Future prediction & estimation of faults occurrences in oil pipelines by using data clustering with time series forecasting. *Journal of Loss Prevention in the Process Industries*, 66, 104203. https://doi.org/10.1016/j.jlp.2020.104203

29. Maheswari, C., Priyanka, E.B., Ibrahim Sherif, IA., Thangavel, S., Ramani, G., (April 2020). Vibration signals-based bearing defects identification through online monitoring using LABVIEW. *Journal Européen des Systèmes Automatisés*, Vol. 53, No. 2, pp. 187-193.

30. Chennippan, Maheswari, Priyanka E. Bhaskaran, Thangavel Subramaniam, Balasubramaniam Meenakshipriya, Kasilingam Krishnamurthy, and Varatharaj Arun Kumar. Design and Experimental Investigations on NOx Emission Control Using FOCDM (Fractional-Order-Based Coefficient Diagram Method)-PIλDµ Controller, *Journal Européen des Systèmes Automatisés*, Vol. 53, No. 5, pp. 695-703. https://doi.org/10.18280/jesa.530512

31. Bhaskaran, P.E., Maheswari, C., Thangavel, S., Ponnibala, M., Kalavathidevi, T. and Sivakumar, N.S., IoT Based monitoring and control of fluid transportation using machine learning. *Computers & Electrical Engineering*, 89, (2021), p.106899. https://doi.org/10.1016/j.compeleceng.2020.106899

32. Ali T, Fatimab N, Biradar RC. A miniaturized multiband reconfigurable fractal slot antenna for GPS/GNSS/Bluetooth/WiMAX/X-band application. *J Electron Commun (AEU)* 2018; 94:234–43.

33. Sharma SK, Shafai L, Jacob N. Investigation of wide-band microstrip slot antenna. *IEEE Trans Antennas Propag* 2004;52(3):865–72.

34. Bhaskaran, P. E., Chennippan, M., & Subramaniam, T. (2020). Future prediction & estimation of faults occurrences in oil pipelines by using data clustering with time series forecasting. *Journal of Loss Prevention in the Process Industries*, 66, 104203.

35. Chen HD. Broadband CPW-fed square slot antennas with a widened tuning stub. *IEEE Trans Antennas Propag* 2003;51(8):1982–6.

36. Sim CYD, Han TY, Liao YJ. A frequency reconfigurable half annular ring slot antenna design. *IEEE Trans Antennas Propag* 2014;62(6):3228–31.

37. Priyanka, E.B. and Thangavel, S., 2020. Influence of Internet of Things (IoT) In Association of Data Mining Towards the Development Smart Cities-A

Review Analysis. *Journal of Engineering Science & Technology Review*, 13(4), pp. 1-21. doi:10.25103/jestr.134.01

38. Priyanka, E.B., Thangavel, S. and Gao, X.Z., 2020. Review analysis on cloud computing based smart grid technology in the oil pipeline sensor network system. *Petroleum Research*. https://doi.org/10.1016/j.ptlrs.2020.10.001

39. Eldek, Elsherbeni, Smith. Rectangular slot antenna with patch stub for ultra-wideband applications and phased array systems. *Progr Electromagn Res, PIER* 2005;53(July):227–37.

40. Maheswari, C., Priyanka, E.B., Thangavel, S., Vignesh, S.R. and Poongodi, C., 2020. Multiple regression analysis for the prediction of extraction efficiency in mining industry with industrial IoT. *Production Engineering*, 14(4), pp. 457-471.

41. Sivanandam SN, Deepa SN. *Principles of soft computing*. 2nd ed. India: Wiley, 2007.

42. Priyanka, E. B., Maheswari, C., & Thangavel, S. A smart-integrated IoT module for intelligent transportation in oil industry. *International Journal of Numerical Modelling: Electronic Networks, Devices and Fields*, e2731.

43. Narayana JL, Krishna KSR, Reddy LP. Design of microstrip antennas using artificial neural networks. In: *International conference on computational intelligence and multimedia applications*, vol. 1; 2007. p. 332–34.

44. Priyanka, E. B., Maheswari, C., Thangavel, S. & Bala, M. P. (2020) Integrating IoT with LQR-PID controller for online surveillance and control of flow and pressure in fluid transportation system. *Journal of Industrial Information Integration*, 17, 100127.

45. Priyanka, E., Maheswari, C., Ponnibala, M., & Thangavel, S. (2019). SCADA Based Remote Monitoring and Control of Pressure & Flow in Fluid Transport System Using IMC-PID Controller. *Advances in Systems Science and Applications*, 19(3), 140-162.

46. Rodrigues EJB, Lins HWC, D'Assuncao AG. Fast and accurate synthesis of electronically reconfigurable annular ring monopole antennas using particle swarm optimization and artificial bee colony algorithms. *IET Microwave Antenna Propagat* 2016:1–8.

47. Priyanka, E. B., S. Thangavel, V. Madhuvishal, S. Tharun, K. V. Raagul, and CS Shiv Krishnan. Application of Integrated IoT Framework to Water Pipeline Transportation System in Smart Cities. In *Intelligence in Big Data Technologies—Beyond the Hype*, pp. 571-579. Springer, Singapore. https://doi.org/10.1007/978-981-15-5285-4_57

48. Sivia JS, Pharwaha APS, Kamal TS. Analysis and design of circular fractal antenna using artificial neural networks. *Progr Electromagn Res B* 2013; 56:251–67.

49. Khan T, De A. Prediction of slot shape and slot size for improving the performance of microstrip antennas using knowledge-based neural networks. In: *International Scholarly Research Notices* vol. 2014, Article ID 957469. Hindawi Publishing Corporation.

50. Bhaskaran, P.E., Subramaniam, T., Kaliyannan, G.V., Palaniappan, S.K. and Rathanasamy, R., 2020. Green Adhesive for Industrial Applications. *Green Adhesives: Preparation, Properties and Applications*, pp. 57-84.
51. Araújo WC, D'Assuncao AG, Mendonca LM. Effect of square slot in microstrip patch antennas using artificial neural networks. *IEEE conference on electromagnetic field computation (CEFC)*, 14th Biennial; May 2010.
52. Guney K, Gultekin SS. Artificial neural networks for resonant frequency calculation of rectangular microstrip antennas with thin and thick substrates. *Int J Infrared Millimeter Waves* 2004;25(9).
53. Priyanka EB, Maheswari C, Thangavel S. Proactive Decision Making Based IoT Framework for an Oil Pipeline Transportation System. In *International conference on Computer Networks, Big data and IoT 2018 Dec 19* (pp. 108-119). Springer, Cham.
54. Thakare VV, Singhal PK. Bandwidth analysis by introducing slots in microstrip antenna design using ANN. *Progr Electromagn Res M* 2009;9:107–22.
55. Youngwook K, Keely S, Ghosh J, Hao L. Application of artificial neural networks to broadband antenna design based on a parametric frequency model. *IEEE Trans Antennas Propag* 2007;55(3).

12

Reconfigurable Antennas

Dr. K Suman

Department of ECE, CBIT, Hyderabad, Telangana

Abstract

In the technologies, the Radio Frequency front end should be conceptual in a natural scenario; reconfigurable antennas have become crucial for the coming generation of wireless communication and systems that are sensible because of their ability to change the radiation characteristics dynamically. They have many advantages such as good isolation, out of band rejection, multifunctional capabilities, low volume, low front end processing efforts without the need for filtering element which made them useful in wireless communications applications such as fourth-generation (4G) and fifth-generation (5G) mobile terminals. Reconfigurable antennas threw a novel challenge to antenna designers and researchers as they can be tuned to any frequency of operation without changing the radiation pattern. For the past thirty years, a lot of improvement was done in the advancement of reconfigurable antennas. This chapter emphasizes the advancements of reconfigurable antennas with basic concepts and gives a few guidelines for future research.

Keywords: Reconfigurable antennas, frequency reconfigurable antennas, polarization reconfigurable antennas, pattern reconfigurable antennas, leaky wave antennas, phase shifters, arrays

12.1 Introduction

Wireless communication systems are moving towards multiple functions of wireless services for different applications. These are used at different times and for different purposes like defense, naval, or domestic purposes. The congestion of the electromagnetic spectrum became one of the reasons for enhancement. To take up this challenge, the upcoming wireless

Email: suman_me192@cbit.ac.in

Prashant Ranjan, Dharmendra Kumar Jhariya, Manoj Gupta, Krishna Kumar, and Pradeep Kumar (eds.)
Next-Generation Antennas: Advances and Challenges, (203–220) © 2021 Scrivener Publishing LLC

communication systems must be cognitive in nature and reconfigurable. The most suitable communication approach is based on the frequency of operation, the direction of the main beam, and different modulation schemes used in the system. To accomplish this, the traditional antennas are replaced by reconfigurable antennas (RA) because of their adaptive radiation characteristics.

They can also be used as control elements that can be programmed with proper feedback to increase the output, noise reduction, and also reduce errors, to avoid obstruction, accumulate energy increase security, lower weakening of the signal caused by fading due to multipath, and also to expand the lifetime of the whole system. Some of the examples of promising applications comprise software-defined radio, cognitive radio, multiple-input multiple-output (MIMO) systems, multifunction wireless devices, and phased arrays with very good performance. A clear understanding of the fundamental characteristics like gain, radiation pattern, frequency of operation, and impedance matching will help in designing a good antenna. When an antenna designer wants to design a reliable antenna these characteristics become a hindrance in the case of the reconfigurable antenna. Reconfiguration can be achieved by redistributing the currents on the antenna or the electromagnetic fields in the antenna's aperture edges [1–4]. The modifications in the operation of the antenna permit its usage in multiple wireless communication applications. With the help of switches or mechanical actuators, antenna design can be reconfigured. Antenna reconfiguration can also be done by varying the following parameters like frequency, polarization, and radiation pattern or by combining all of them. The classification of RAs (shown in Figure 12.1) mostly depends on the above three parameters and is further divided into four groups such as electrical devices like RF-MEMS, PIN diodes, and Varactor

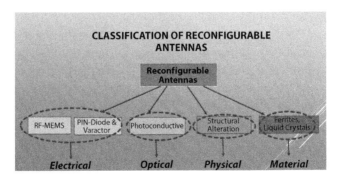

Figure 12.1 Classification of reconfigurable antennas.

diodes; optical devices like photoconductive devices; by physically altering the structure; and by material change like ferrites, liquid crystal, etc.

12.2   Reconfigurability of Antenna

A reconfigurable antenna is defined as one that has tunable elemental characteristics which include frequency of operation, impedance, bandwidth, radiation pattern, and polarization. The main aim is to modify the characteristics without affecting the others. As antennas are mainly electromechanical elements, the characteristics in which they are altered are electrical, mechanical, or electromechanical. The antenna designers must be capable of designing sophisticated antennas that can be cognitive and easily adaptable to varying environmental situations. These antennas should overcome failures and drawbacks and should be fast enough to respond to the new developments in technology. The applications for the assimilation of extremely flexible, dependable, and proficient reconfigurable antennas include Cognitive radio, Multiple-input multiple-output (MIMO) channels, On-body networks, Satellites, and Space communication platforms.

12.2.1   Frequency Reconfigurable Antennas (FRAs)

Frequency reconfigurable antennas can change their frequency of operation. They are most helpful in situations where many communication systems congregate since the multiple antennas necessary can be exchanged by a single reconfigurable antenna. Frequency reconfiguration is obtained by a physical or electrical variation to the antenna dimensions using RF-switches, impedance loading, or tunable materials. They have the reconfiguration of the resonant frequency by changing the structure, while the radiation patterns and polarization remain unchanged. Generally, a definite radio tool has to take care of many services on a wide choice of frequencies. Like all electronic gadgets are supported by different values like Wireless Local Area Network (WLAN), Worldwide Interoperability for 3G, 4G, Microwave Access (WiMAX), Bluetooth, and GPS. However, antennas are necessary to cover up multiple operating bands. But if a particular band is to be selected, then reconfigurable antennas are preferred over a wide band and multiband antennas because RAs can be made more compact, and also they present rejection in noise in the band, which are not used, and reduce the number of filters required at the front end. Frequency RAs can be classified into two types; frequency continuous tuning and discrete tuning.

12.2.1.1 Continuous Tuning

Continuous frequency tuning can be achieved by using Varactor diodes. The edges through which the patch antenna radiates is connected to Varactor diodes to comprehend frequency sharpness [5, 6]. The required frequency tuning ratio can be produced by varying the well-organized electrical dimension of the patch which is achieved by varying the bias voltage of the Varactor diodes. A differentially fed frequency-response Microstrip patch antenna was proposed [7] as shown in Figure 12.2.

The antenna is capable of achieving a 2.0 frequency tuning ratio on the Microstrip patches by attaching three pairs of Varactor diodes. These antennas act as better devices for different designs of frequency RA. Generally, Varactor diodes are used to vary the dimension of the slot which helps in regulating the frequency of the antenna. The slot frequency RAs can provide a broader regulation ratio around 3.52 but at the cost of gain and efficiency. The frequency reconfiguration modifies the electrical length of the dipole with the help of Varactor diodes or switches. A good frequency reconfigurable design can provide better selectivity of the frequency and also lowers the undesirable effect of co-site intervention and congestion.

12.2.1.2 Discrete Tuning

Discrete frequency tuning can be achieved through PIN diode and MEMS switches. A microstrip patch antenna with a compressed frequency response can be considered as the best example [8]. Using PIN diodes the

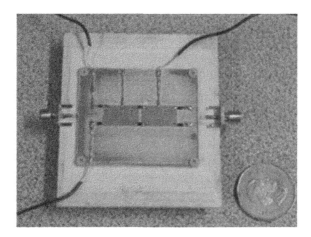

Figure 12.2 Microstrip patch antenna with differentially fed frequency-agile [7].

central patch is coupled to four dissimilar peripheral elements. All of them are designed according to the specifications and operate at a certain frequency, self-sufficient from the others. For a broad range of frequencies between 0.8 to 3.0 GHz, the patch antenna can attain the whole of 24 different states.

12.3   Polarization Reconfigurable Antenna (RA)

By varying the angles of linear polarizations, or exchanging among linear and circular polarizations, or between left-hand circular polarization (LHCP) and right-hand circular polarization (RHCP), polarization reconfiguration can be achieved. Polarization diversity is used to enhance the competence of multiple-input-multiple-output (MIMO) system [9, 10]. Antennas that employ this can be used to offer polarization diversity to reduce the signal fading in multipath propagation conditions. The most important challenge faced in designing an antenna is achieving polarization agility. This can be done without any alteration in the antenna input impedance characteristics. Hence it is a challenge for the designers to devise a polarization reconfigurable antenna that can simply change among linear and circular polarization as it becomes complicated in achieving impedance matching for these two polarizations. The two degenerate orthogonal linear modes are used to generate circular polarization (CP), but its input impedance varies from the one that is used to generate a linear polarization (LP). In comparison with a single band antenna, the interdependence of polarization characteristic of CP and its frequency response are much prominent. Because it is difficult to design a dual-band polarization reconfigurable antenna. Hence it becomes a challenge for the designers to sustain the state of polarization that belongs to the dual/multiple bands by maintaining the frequency stability. In the section below, different methods are presented that switch among linear, circular, and double-band polarization reconfigurable antennas.

12.3.1   Polarization RA with Single Band

To exchange between circular and linear polarization, many interesting polarization reconfigurable antenna designs are proposed. To operate in LP and CP modes a slot antenna with square rings and four-pin diodes was designed by Dorsey et al. [11]. In combination with four-pin diodes with small bandwidth on a compact square patch to create CP and LP, radiation was proposed by Sung et al. [12]. A Microstrip circular patch antenna

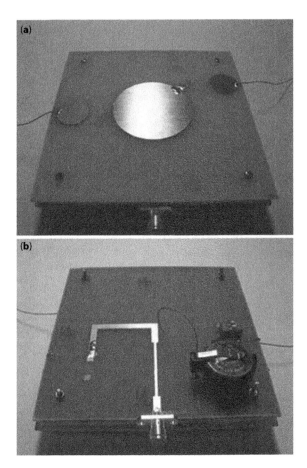

Figure 12.3 Reconfigurable antenna with circular patch: (a) Top layer; (b) Bottom layer [12].

with coupled ring-slot devised. It can change among linear and circular polarizations. This covers the bandwidth of 2.2%. The figure of polarization reconfigurable antenna with the circular path is shown in Figure 12.3.

12.3.2   Dual-Band Polarization RA

In the present scenario, a dual-band operation is very important in the WLAN devices. A frequency band operating at 2.4 and 5.8 GHz band is preferred for polarization RA. A polarization reconfigurable antenna with single aperture fed was devised by Qin *et al.* [13] to meet this requirement. In both 2.4 and 5.8 GHz bands, it continuously changes between horizontal, vertical, and 450 linear polarizations.

12.3.3 Pattern Reconfigurable Antenna (RA)

Pattern Reconfigurable Antenna can minimize interference by varying the null positions so that it saves energy. By adjusting the main beam, a large area can be covered which intends to direct the signal towards the required number of intended users. These RAs have the potential to modify the shape of the main beam or may scan the main beam. Further, the overall system ability is enhanced by using multiple-input-multiple-output (MIMO) systems. The antenna structure helps in finding the radiation pattern with the flow of current through the structure. The reconfigurability is accomplished by changing the current distribution. Further, the frequency characteristics should not be changed for various radiation patterns of the antenna. Without changing the operating frequency it is very difficult to get pattern reconfigurability. Various procedures have been implemented to get through this challenge. The change in antenna structure such as parasitically coupled antennas or reflector antennas is one of them. This helps in achieving an independent input feed port for the reconfigured structure. This in turn makes the frequency characteristics to be steady. Some additional structures are also used to balance the variations in the input impedance of the antenna.

12.3.4 Main-Beam Shape

The wideband circular patch antenna is used as an example that can be switched among boresight and conical radiation patterns. The switching can be done using L-probe feeds [14]. For boresight radiation, TM_{11} mode is stimulated and for conical radiation, TM_{01} is stimulated by the two feeds. The conical pattern mode's resonant frequency is reduced by attaching the patch with shorting posts. This helps in increasing the frequency of operation of the modes. Reconfiguration of the radiation pattern is done by using an integrated matching network with sufficient switches. It was also devised that broadband tunable Coplanar Waveguide (CPW)-to-Slot line transition feed for a particular pattern can be used [15]. Reconfiguration of the feed mode among left side slotline (LS) mode, right side slotline (RS) mode, and Coplanar Waveguide mode is done by PIN diodes. The left side slotline (LS) mode and the right side lot line (RS) mode stimulate both the patterns along the axis and their radiating beams are intended in reverse order.

12.3.5 Main Beam Scanning

To guide the major beam to already defined directions, several antennas have been proposed. By using PIN diodes or any other electronic switches, to activate one or several elements. A four-element L-shaped antenna array was proposed by Lai *et al.* [16]. It can easily accomplish beam direction-finding upon 3700 in the plane of azimuth with a gain of approximately 0.6-2.0 dB. Substantial advances have been made in creating the beam-steering pattern RA. Their application is limited because of low gains. A reconfigurable Quasi-Yagi dipole with a high gain beam switching pattern was proposed by Qin *et al.* [17]. It can direct the E-plane major beam in either 300, -300, or 00 with a gain between 6.5 and 10dBi.

12.4   Compound Reconfigurable Antennas (RAs)

The aim of designing a reconfigurable antenna is to achieve the ability to alter the radiation pattern, polarization, and operating frequency. This can easily be achieved in a compound reconfigurable antenna. The Compound RAs are more flexible and diverse than any single characteristic antenna and hence play a very important role in wireless communication systems.

Radiation pattern reconfigurability and combined frequency can be easily achieved [18]. Figure 12.4 shows the annular slot antenna configuration. The direction of the null can be changed by proper loading the PIN diodes from corner to corner of the slot at definite locations (Figure 12.4a). The frequency of operation of the antenna can be altered by redesigning the network as shown in Figure 12.4b.

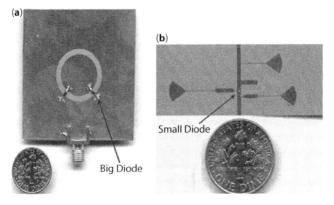

Figure 12.4 (a) Annular slot antenna - Front side, (b) Impedance matching network - Backside [18].

12.5   Reconfigurable Leaky Wave Antennas

These antennas are considered to be the transmission lines that slowly give away their energy into free space. With the help of the phase constant β, the radiated energy θ can be found for the leaky wave throughout the distance of the transmission line [19]. The output of the leaky-wave antenna is shown in Figure 12.5.

The beamwidth is illuminated by controlling the leakage rate α and is found from the length of the leaky aperture. The advantage of leaky wave antennas over reflector and lens antennas is that it can perform well without protruding feed and can also maintain a minimum profile. When they are compared with arrays a network with the feed is not required, because this feed sometimes becomes lossy for big arrays. The beam direction can be steered by sweeping the operating frequency, as the phase constant of the leaky-wave antenna depends on the operating frequency.

The wireless communication industry and its standards have grown at a fast pace in the last decade. The spectrum regulators have defined a frequency band on which the wireless communication works, and thus the inbuilt frequency scanning nature of leaky-wave antenna is inadequate. This gives rise to the progress of pattern reconfigurable (beam-steering) leaky-wave antennas that exploit the complex nature of the propagation channel. Further, they have a relatively narrow bandwidth. But Frequency Reconfigurable (tunable) Leaky-wave antenna can still solve the purpose and become very useful in many applications. For effective channelization, fixed frequency operation is required. This limits their application in modern communication systems.

However, considerable attempts are made to develop frequency-independent leaky-wave antennas. PIN diodes are used as switches and transformed the radiation angle electrically by regulating the guided wavelength [20]. Only two discrete angles were present as the diodes have only two

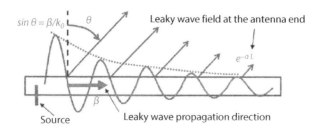

Figure 12.5 Output of leaky-wave antenna.

states namely biased and unbiased. The DC magnetic field is tuned and the radiation angle is using Ferrite slab to build the leaky-wave antenna which is scannable magnetically [21].

12.6 Reconfigurable Antennas - Applications in Wireless Communication

This section discusses the dissimilar practical wireless communication systems which use reconfigurable antennas.

12.6.1 Reconfigurable Antennas - MIMO Communication Systems

Several antennas are erected on either side of the communication channel in the MIMO communication system. Multiple signals can be transmitted and received at the same time. It will overcome the multipath and fading and also increases the capacity of the channel and also spectral efficiency. The reconfiguration in polarization and radiation patterns improves the overall channel capacity [3, 9, 22–25]. An Electronically Steerable Passive Array Radiator (ESPAR) serves as the best example for an antenna that can be used for channels with MIMO which is shown in Figure 12.6. It contains a driven monopole situated in the center enclosed by a ring of six evenly spaced parasitic monopoles.

Each component is terminated with a load that contains a 0.3nH inductor and a PIN diode. The load provides the required reactance when a monopole is used as a reflector when it is capacitive or a director when it is inductive [26]. Among these, three elements are arranged as reflectors and

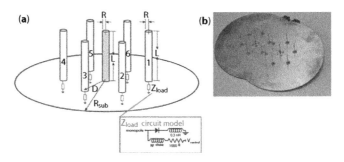

Figure 12.6 ESPAR for MIMO communications [26].

the rest of them as directors then they will generate a directive beam in the required direction. ESPAR can produce six different patterns with beam angles 600 away from each other in azimuth if it is circularly configured. The design is optimized to maximize pattern flexibility.

12.6.2   Reconfigurable Antennas - Mobile Terminals

Reconfigurable antennas can be used for mobile terminals. They facilitate them to switch over from one operational protocol to another easily and in very less duration of time. Mobile terminals like smartphones, GPS Receivers, Laptops, or other compact devices adopt the Multiband antennas so that they can perform using diversified wireless protocols. Several experiments are done with the help of a reconfigurable antenna to determine the advantages of multiple antennas used for mobile terminals. For example, when using a reconfigurable antenna or a multiband antenna, the GPS receiver is subjected to blocking of the signal. In such a case the multiband antenna can operate with GPS and WLAN frequencies and the reconfigurable antenna switches for its operation from GPS to WLAN. There is no degradation of the signal noticed when the reconfigurable antenna is used but the multiband antenna shows a reduced band rejection. So this needs extra filtering to obtain better noise rejection. With fundamental passband characteristics and exceptional out of band rejection in the absence of filters, the Frequency reconfigurable antennas are widely used. Their resonances being extremely far away from the working band, they can be easily filtered at a lower cost [27].

12.6.3   Reconfigurable Antennas for Cognitive Radio Applications

The combination of Cognitive radio with reconfigurable antenna improves the communication efficiency effectively [28]. A cognitive radio continuously monitors the medium and searches for the left out gaps in the spectrum. Once the spectrum is recognized a reconfigurable antenna activates and modifies itself to broadcast. Thus cognitive radio can recognize the changes in a communication system and adjust accordingly. A wideband antenna or a narrow band reconfigurable antenna can be used for sensing the channel. In the communication process, dynamically revealed that white spaces in the spectrum must be well managed by a reconfigurable antenna. The operation of the cognitive radio is well explained by a cyclic diagram shown in Figure 12.7.

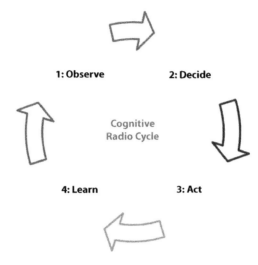

Figure 12.7 The cognitive radio cycle.

As shown in the figure, the operation of the cycle is as follows:

a) ˙ The channel activity is first observed by sensing (wide-band) antenna.
b) The cognitive processor makes a decision or decides as to which part of the spectrum is most appropriate for communication.
c) The communicating antenna is activated by the processor to get the required mode of communication.
d) Finally the cognition is achieved by the processor by getting trained from earlier channel activities.

The selected mode of communication is accomplished by the device used for cognitive radio. It can determine and self-reconfigure the hardware used optimally.

12.6.4   Reconfigurable Antennas - MIMO-Based Cognitive Radio Applications

Several antenna arrangements have been recommended for cognitive radio applications [29]. But it becomes difficult to communicate over idle frequency gaps and also to add cognition to the systems. This, however, does not solve the issues in the spectrum. The antenna system should be designed in such a way that it should be able to address the problems of the

Figure 12.8 MIMO-based reconfigurable filtenna [29].

spectrum such as multipath or fading in cognitive radio and also solve any other issues in the spectrum. A MIMO-related antenna is recommended for applications in cognitive radio as shown in Figure 12.8. This improves spectrum usage frequency, combats fading, and ensures reliable communication connecting the end users. It contains two sensing antennas along with a pair of reconfigurable "filtennas" loaded and combinable on the identical substrate. To make it compatible with their frequencies in which they are operated, the reconfigurable antennas choose PIN diodes.

12.6.5   Reconfigurable Antennas - WLAN Band Rejection

Interference is the most common problem encountered between UWB antennas and WLAN signals for communication networks. It is because of the increased traffic the allocation of WLAN to a large number of mediums in the highest operational band (5.15–5.825 GHz) and UWB antennas, have large bandwidth over 802.11n usage, Federal Communications Commission [30, 31]. By using WLAN with USB antennas, the problem of interference can be resolved.

12.6.6   Reconfigurable Antennas - Wireless Sensing

Reconfigurable antennas are used in wireless sensing applications like habitat monitoring, industrial process control, implantable medical telemetries, and battlefield surveillance to mention a few of them.

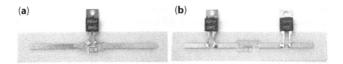

Figure 12.9 Proposed reconfigurable sensing antenna.

Figure 12.9 shows a proposed reconfigurable antenna for wireless sensing. The printed dipole is used as an antenna attached to the T-matching network. The impedance of the antenna is combined with an RFID Integrated Circuit. On an FR4 substrate, the system is printed. By using the antenna sensitivity and gain the read range of the tag antenna can be determined. The impedance of the RFID microchip varies with the frequency and the input power. By joining the thermal switches in series and parallel these reconfigurable antennas monitor the changes in temperature of the surroundings. However, the change in the temperature significantly affects the antenna performance. The change is noticed as the temperature goes beyond a particular threshold value.

12.6.7   Reconfigurable Antennas - Terahertz (THz) Communication Applications

In recent decades, the use of TeraHertz (1012) frequency in communications has laid a new pavement for many antenna researchers. This was made possible by the advent of graphene as a design material for the antenna. The plasmonic modes are supported by the graphene and help in reducing the size or miniaturization of the devices that are used for communication and sensing purposes [32]. The graphene used in antennas provides high miniaturization, high efficiency, better matching, and inherent reconfiguration capabilities. The antennas that use plasmonic modes on graphene sheets can combine Electromagnetic energy from very small Terahertz sources to free space.

12.6.8   Reconfigurable Antennas - Millimeter-Wave Communication Applications

Millimeter (mm) wave frequencies between 30 and 300 GHz are used for communication purposes because of their benefits. Antenna system designers focus on its wide bandwidths useful for high-speed data transmission and also video transmission. The mm waves can travel only for a

Figure 12.10 "mm" wave antenna [34].

very short distance. These waves cannot pass through solid materials easily and their performance is influenced by ecological factors such as rain and for federal Communications Commission [33]. Several RAs are developed to operate at "mm" wave frequencies. An example of an "mm" wave antenna is shown in Figure 12.10.

This antenna is augmented by a corporate feeding network and has 16 distinct patches. These patches are arranged in the shape of a semicircle. These patches are connected to the waveguides and are exactly coupled to the MMIC power amplifiers. Lenses are then connected to the antenna system and these lenses are illuminated by the sources of the patch. An inhomogeneous lens with an infinity focus point called a Luneburg lens is used. Once the lens is illuminated by these patches, the other patches are loaded with 50Ω. Reconfigurable antennas can also use the liquid crystal for mm waves.

12.7   Optimization, Control, and Modeling of Reconfigurable Antennas

To meet the varying communication channel requirements, reconfigurable antennas are used because they can transit from one state to another easily. To this aspect, designers have proposed various optimization algorithms to smooth the transformations among different states of a reconfigurable antenna, such as particle swarm optimization, simulated annealing, genetic algorithm, ant colony optimization, self-organized maps, the cross-entropy method, and the self-adaptive-induced mutation algorithm. The negativity of any reconfiguration technique on the performance of the antenna is minimized by all these algorithms. In contrast to these optimization techniques, these algorithms cannot be departed from others because of the best fit, before selecting a precise reconfigurable design. Another aspect to be considered while designing a reconfigurable antenna is its complexity. This increases costs and losses. To reduce the complexity, several approaches have been anticipated without disturbing the reliability of the antenna system. Generally, these are based on diverse techniques or models that are previously existing. The learning state selection approach is one such technique that determines the antenna behavior for various reconfigurations. Nowadays, neural networks are used to study the behavior of the antenna that can help to create various antenna states based on prior learning. However, the use of neural networks has reduced computational complexity.

12.8   Conclusions

In this chapter, queries interrelated to the design of reconfigurable antennas are addressed. The need to design reconfigurable antenna structure is discussed in this chapter and also in the applications where they are potential solutions. They form the most important element of the advanced communication system. A dynamic communication system that relies on antenna structure is required to cope with the emerging trends. Future work in this area of the reconfigurable antenna includes their design that can distribute the full reconfiguration of the antenna. Such a RA is functional to numerous wireless communication systems. Control over the frequency of operation, its polarization, and the radiation pattern is extremely challenging for the tough relationship between the antenna's frequency reaction and its radiation characteristics. To realize the full reconfiguration of an antenna, novel and useful techniques are required to come out of the linkage. The future work can also be extended to RF and also to the signal processing in the physical layer. As antennas form only a part of the entire communication system, the remaining parts in the system must be capable of reconfiguring their features so that they can be utilized in a diversified manner. Therefore reconfigurable antennas play a very important role in enhancing the next-generation communication systems considerably.

References

1. Balanis CA (2011) *Modern antenna handbook*. Wiley-Interscience, Hoboken, NJ, USA.
2. Bernhard JT, Volakis JL (2007) *Antenna engineering handbook*, 4th ed. McGraw-Hill, New York.
3. Christodoulou CG, Tawk Y, Lane SA, Erwin SR (2012) Reconfigurable antennas for wireless and space applications. *Proc IEEE* 100(7):2250–2261.
4. Costantine J, Tawk Y, Christodoulou CG (2013c) *Design of reconfigurable antennas using graph models*. Morgan and Claypool, San Rafael, CA, USA.
5. Bhartia P, Bahl IJ (1982) Frequency agile Microstrip antennas. *Microw J* :1982, 25, 67–70.
6. Waterhouse R, Shuley N (1994) Full characterization of varactor-loaded, probe-fed, rectangular, Microstrip patch antennas. *IEE Proc Microw Antennas Propag* 141:367–373.
7. Hum SV, Xiong HY (2010) Analysis and design of a differentially-fed frequency agile Microstrip patch antenna. *IEEE Trans Antennas Propag* 58:3122–3130.
8. Genovesi S, Candia AD, Monorchio A (2014) Compact and low profile frequency agile antenna for multistandard wireless communication systems. *IEEE Trans Antennas Propag* 62:1019–1026.
9. Qin PY, Guo YJ, Liang CH (2010a) Effect of antenna polarization diversity on MIMO system capacity. *IEEE Antennas WirelPropag Lett* 9:1092–1095.
10. Qin PY, Weily AR, Guo YJ, Bird TS, Liang CH (2010b) Frequency reconfigurable quasi-Yagi folded dipole antenna. *IEEE Trans Antennas Propag* 58:2742–2747.
11. Dorsey WM, Zaghloul AI (2009) Perturbed square-ring slot antenna with reconfigurable polarization. *IEEE Antennas Wirel Propag Lett* 8:603–606.
12. Sung YJ, Jang TU, Kim YS (2004) A reconfigurable Microstrip antenna for switchable polarization. *IEEE Microw Wirel Compon Lett* 14:534–536.
13. Qin PY, Guo YJ, Ding C (2013a) A dual-band polarization reconfigurable antenna for WLAN systems. *IEEE Trans Antennas Propag* 61:5706–5713.
14. Yang SLS, Luk KM (2006) Design a wide-band L-probe patch antenna for pattern reconfigurable or diversity applications. *IEEE Trans Antennas Propag* 54:433–438.
15. Wu SJ, Ma TG (2008) Awideband slotted bow-tie antenna with reconfigurable CPW- to-slotline transition for pattern diversity. *IEEE Trans Antennas Propag* 56:327–334.
16. Lai MI, Wu TY, Wang JC, Wang CH, Jeng S (2008) Compact switched-beam antenna employing a four element slot antenna array for digital home applications. *IEEE Trans Antennas Propag* 56:2929–2936.
17. Qin PY, Guo YJ, Ding C (2013b) A beaming steering pattern reconfigurable antenna. *IEEE Trans Antennas Propag* 61:4891–4899.
18. Nikolaou S *et al* (2006) Pattern and frequency reconfigurable annular slot antenna using PIN diodes. *IEEE Trans Antennas Propag* 54:439–448.

19. Goldstone L, Oliner AA (1959) Leaky-wave antennas I: rectangular waveguides. *IRE Trans Antennas Propag* 7:307–319
20. Horn RE, Jacobs H, Freibergs E, Klohn KL (1980) Electronic modulated beam steerable silicon waveguide array antenna. *IEEE Trans Microw Theory Tech* 28:647–653.
21. Maheri H, Tsutsumi M, Kumagi N (1988) Experimental studies of magnetically scannable leaky-wave antennas having a corrugated ferrite slab/dielectric layer structure. *IEEE Trans Antennas Propag* 36:911–917.
22. Cetiner BA, Jafarkhani H, Qian JY, Yoo HJ, Grau A, De Flaviis F (2004) Multifunctional reconfigurable MEMS integrated antennas for adaptive MIMO systems. *IEEE Commun Mag* 42(12):62–70.
23. Grau A, Romeu J, Lee M, Blanch S, Jofre L, De Flaviis F (2010) A dual linearly polarized MEMS reconfigurable antenna for narrowband MIMO communication systems. *IEEE Trans Antennas Propag* 58(1):4–16.
24. Li Z, Du Z, Gong K (2009) Compact reconfigurable antenna array for adaptive MIMO systems. IEEE *Antennas WirelPropag Lett* 8:1317–1321.
25. Piazza D, Kirsch NJ, Forenza A, Heath RW, Dandekar KR (2008) Design and evaluation of a reconfigurable antenna array for MIMO systems. *IEEE Trans Antennas Propag* 56(3):869–881.
26. Zhou Y, Adve RS, Hum SV (2014) Design and evaluation of pattern reconfigurable antennas for MIMO applications. *IEEE Trans Antennas Propag* 62(3):1084–1092.
27. Yang S, Zhang C, Pan HK, Fathy AE, Nair VK (2009) Frequency-reconfigurable antennas for multi radio wireless platforms. *IEEE Microw Mag* 10(1):66–83.
28. Tawk Y, Costantine J, Christodoulou CG (2014a) Cognitive radio antenna functionalities: a tutorial. *IEEE Antennas Propag Mag* 56(1):231–243.
29. Tawk Y, Costantine J, Christodoulou CG (2014b) Reconfigurable filtennas and MIMO in cognitive radio applications. *IEEE Trans, Antennas Propag* 62(3):1074–1083.
30. Anagnostou DE, Chryssomallis MT, Braaten B, Ebel JL, Sepulveda N (2014) Reconfigurable UWB antenna with RF-MEMS for on-demand WLAN rejection. *IEEE Trans Antennas Propag* 62(2):602–608.
31. Federal Communications Commission (2002) First report and order revision of part of the commission's rule regarding ultra-wideband transmission systems. FCC 02–48
32. Tamagnone M, Gomez Diaz JS, Perruisseau-Carrier J, Mosig JR (2013) High-impedance frequency-agile THz dipole antennas using graphene. In: *7th European conference on antennas and propagation,Gothenburg, Sweden*, pp 533–536.
33. Federal Communications Commission (1997) Millimeter wave propagation: spectrum management implications. FCC Bulletin 70.
34. Lafond O, Himdi M, Merlet H, Lebars P (2013), An active reconfigurable antenna at 60 GHz based on plate inhomogeneous lens and feeders. *IEEE Trans Antennas Propag* 61(4):1672–1678.

13

Design of Compact Ultra-Wideband (UWB) Antennas for Microwave Imaging Applications

Dr. J. Vijayalakshmi* and Dr. V. Dinesh

*Department of ECE, Kongu Engineering College, Perundurai,
Erode, Tamil Nadu, India*

Abstract

In spite of the current tremendous technological improvement in the medical field, there is still a need for more research on the early detection of cancer. Yearly, more people are diagnosed with cancer. At present, there are many imaging modalities used by the clinician: X-ray, Magnetic Resonance Imaging (MRI) scan, Computed Tomography (CT) scan, mammography, Positron Emission Tomography (PET) scan, optical imaging, and Ultrasound. Due to current limitations on the above imaging methods, some alternative investigations and research tools are encouraged for the early detection of cancer. Ultra-wideband (UWB) antenna is remarkably an appropriate antenna in the detection of cancer.

In this chapter, an Ultra-wideband miniaturized antenna for microwave imaging is designed. UWB antenna is operated for the frequency range of 2.4-10.6GHz. The FR4 substrate with a dielectric constant of 4.2 and height of h = 1.6mm is preferred for the design of the antenna at the UWB frequency range. The Federal Communications Commission (FCC) in 2002 [1] has stated the frequency range for UWB signal from 3.1 GHz to 10.6 GHz. To design the multiple antennae to get operated for multiple frequencies, space occupancy is the major problem in many cases. To resolve this problem, many researchers have taken an effort to design different types of antennas like slot antenna, I-shaped radiator, reconfigurable antenna, and metamaterial antennas. These antennas are suitable for various medical applications like blood glucose testing, heart rate detection, and breast cancer detection [2].

Keywords: UWB, monopole antenna, defected ground structure, microwave imaging, bowtie antenna

*Corresponding author: vijaya.jagadeesh09@gmail.com

Prashant Ranjan, Dharmendra Kumar Jhariya, Manoj Gupta, Krishna Kumar, and Pradeep Kumar (eds.)
Next-Generation Antennas: Advances and Challenges, (221–250) © 2021 Scrivener Publishing LLC

13.1   Introduction

Especially in the UWB applications, planar monopole antenna plays a vital role owing to its advantages like miniaturization, cost-effectiveness on fabrication, high flexibility to integrate with the transceiver system. Presently, the ionization radiation effect of Microwave Imaging (MI) gets altered from the existing diagnostic methods by the high radiation effects. Microwave imaging is the active non-invasive method in which electromagnetic signals penetrate inside the human tissues. Hence, the depth and direction of the tumour present inside the body can be identified easily. Electromagnetic signals do not produce thermal heat to the tissues. The microwave imaging method is classified as Microwave Tomography and UWB radar detection technique. The non-invasive-based UWB radar detection technique is the most effective methodology for detecting brain stroke, brain tumour, breast cancer, and lung cancer.

13.1.1   Ultra-Wideband Antennas (UWB)

UWB (Ultra-Wideband) is a radio communication technology that uses very low energy pulses and it is proposed for both short-range signal transmission and also for high bandwidth range in terms of GHz frequency. In the UWB communications, data transmission does not interfere with other conventional band systems in the resonant frequency on a similar band of frequency. It is very high speed and is applicable for the WLAN, HiperLAN wireless technologies. On the 14th February 2002, Report and Order from the FCC (Federal Communications Commission) authorized the usage of the unlicensed UWB of the range from 3.1 to 10.6 GHz for commercial applications. The approved FCC power spectral density emission limit for UWB transmitter is -41.3 dBm/MHz.

In 1893, Hertz experimented on wireless communication using UWB. The wideband pulse waveforms are used as the main technique after Hertz's first electromagnetic experiment. In the technology development of communications, wideband pulses shifted to narrowband sinusoidal waveforms. In the 1960s the generation of short pulses were energized and new investigations were made in the area of impulse radio. These challenges were stimulated in the design of the UWB antenna for radar communication. UWB signals are measured as impulse, carrier-free, time-domain signal which is relative to the bandwidth of radar or radio signals by the US Department of Science. The typical requirement is to design high data range transmission with a short range of transmission systems that is satisfied by

UWB systems. Later, the FCC approved the first report for validating the usage of UWB systems for commercial applications. FCC (2003) states that the UWB antenna is designed if its bandwidth is greater than 500MHz [8]. The large bandwidth of the signal between 3.1 to 10.6GHz, is determined by UWB systems which have the following benefits: minimum power consumption, high data rate transmission, higher resolution, implementation cost is lower, penetration in obstacles, interference resistance, co-existence with narrowband systems, etc. [10]. These advantages in UWB systems help for a broad range of applications in communications, radar, and imaging. UWB systems have a short-range of pulses of around 10 to 10000ps that is applicable for a broad range of frequencies. UWB technology has the skill to send a huge amount of information with restricted interference signals. It also tends to compete in jamming the signals by their short pulse transmission [11].

Various types of UWB antennas are investigated for the microwave imaging system. Stacked patch antenna, wide slot antenna, tapered slot, bowtie antennas, and so on are suitable antennas for this applications. Lastly, the small-size planar monopole antenna satisfies the requirement of microwave imaging application. Besides this, radar-based microwave imaging uses an Ultra-Wideband (UWB) pulse that can be suitable for all frequency ranges. Antenna designed under low-frequency band used to find a depth of penetration while the high-frequency band ensures the resolution of the resulting images. Hence, the small-size cancer cells and in-depth placed cancer cells can be detected based on the examination results from the low-frequency and high-frequency band antennas.

UWB antenna has major characteristics like gain, bandwidth, radiation pattern, beamwidth, and impedance of an antenna. In the case of early detection of breast cancer, UWB-based Radar-based Microwave Imaging (MI) technique gives better results.

The following are the advantages of Radar-based MI techniques:

a) Implementation cost is low
b) Radiation effect on the body is harmless
c) Detection of tumour accuracy is high
d) The pleasure of the patient on using the devices is better compared to the currently used diagnostic methods.

Challenges to be considered in the usage of UWB signals as

a) Hardware & antenna design
b) Imaging algorithm.

The antenna is utilized to radiate the ultra-wideband signal to obtain the highest resolution in the image. This transceiver antenna has the largest fractional bandwidth with low sidelobes and minimum mutual coupling between two antennas in a multi-static radar approach. The radar approach to breast cancer detection is capable of transmitting UWB signals.

13.2   Microwave Imaging

An alternative method to the existing diagnostic methodology in cancer detection like X-ray, MRI, CT scan, PET scan, and other methods is Microwave Imaging (MI). The existing methodology is injurious to the human body due to their radiation effects. Some of the limitations faced in the currently used diagnosis methods are concentrated in the proposed method, like increasing the resolution of the results, cost-effective and more accuracy while using in all types of soft tissues in the human body. The Scattering principle is used behind the Microwave imaging method, in which the scattering wave or reflected wave from the distinction in dielectric properties between the benign tissue and the cancer cells of the human breast. In this technique, the differential level of the water content of the presence and absence of tumour in the soft tissues is identified. This technique transmits the short pulses of low-power microwave sent by the UWB antennas into the human brain and backscattered energy through the position of an antenna and three-dimensional images from the signal collected. The dielectric properties of tumour tissue from the normal tissue are due to larger water content in the tissues. The microwave imaging method uses a UWB antenna.

Microwave imaging for the brain can be classified into three types:

a)   Passive method
b)   Hybrid method
c)   Active method

The passive method of microwave imaging is based on discriminating the healthy and malignant tissue by increasing the tumour temperature in microwave radiometry. For diagnosis and analysis of brain tumours, the level of temperature measurement is taken into account.

In the hybrid method, the ultrasonic and microwave radiation techniques are employed. The electrical conductivity of the tumour cells is greater than the normal tissue, hence the microwave energy is absorbed by the tumour and ultrasonic transducers are used for heating the tumour and those images are expanded and can be used for detection.

In the active method, microwave signals are radiated into the brain for detecting the existence of a tumour. The reconstruction method is used in the backscattered energy which shows the significance between different grade levels of brain tumour. Active method is subdivided into two approaches, tomography and radar-based approaches.

a)   In the tomography method, a single transmitter is used for electromagnetic radiation into the brain and several antennas are placed around the brain to receive the scattered energy. The transmitter can be moved around various positions, and the data acquired can be used for further processing in a two-dimensional or three-dimensional image of the brain tumour.

b)   In the radar-based microwave imaging approach, the short-pulsed signal is transmitted from a single UWB antenna into the brain. A monostatic or bistatic approach can be used in this technique. In the monostatic approach, the same antenna can transmit the energy, and backscattered waves are received from the same antenna. This process will be continued for various locations around the brain. In the bistatic approach, the array of antennas is used around the brain, every antenna can be acted as transmitter and receiver, same backscattered energy is received by the receiver antenna. The scattered signal is returned from the human body and this may significantly disturb the response to identify the actual tumour location. The travel time of the signals at various locations are recorded and computed. This setup does not require any complicated image reconstruction algorithm of any other radar-based methodology in microwave tomography in microwave imaging method.

Radar-based microwave imaging focuses only on imaging on the tumour and not on the whole phantom. Hence signal processing methods can be easily applied for image resolution stages than in microwave tomography. Signal pulses involve both low and high frequencies. The low-frequency signal is for sufficient depth of penetration while the high frequency involves the adequate resolution of the ensuing images. Therefore this system is helpful for deeply buried and smaller-size tumour. The radar-based imaging method works based on the principle of Ground Penetrating Radar (GPR) in cancer detection. Identification in the dielectric difference in healthy and tumour tissue is from backscattered signal reflects from the location of the tumour. To identify the deeply buried and also the smaller size of the tumour by examining the results obtained signal reflection from the deeply penetrated signal. This signal is transmitted by using microwave tomography and radar-based microwave imaging method [24].

13.3   Antenna Design Implementation

13.3.1   Design of Reflector-Based Antipodal Bowtie Antenna

The antenna parameters of size miniaturization, UWB fractional bandwidth, gain and directivity improvement can be achieved by the design of UWB ground reflector-based bowtie antenna in the antipodal configuration. The stepped feed is designed for multiband frequency resonance over the UWB band and improves the antenna gain. To increase the VSWR bandwidth of an antenna, the design of a triangular-shaped or round-edged wing-like structure can be used as a bowtie patch antenna. The width of the bowtie patch antenna gets increased results in an improvement in the bandwidth of an antenna without any loss in its performance characteristics. The improved directivity characteristics of bowtie patch antenna are most adapted for radar applications such as blind-spot detection in autonomous vehicles and health monitoring systems. Vijayalakshmi *et al.* [20] have proposed that compact size overlay patch antenna design for UWB frequency range with defected ground structure has achieved better gain and highly directional results. Dinesh *et al.* [22] have described that staircase steps in the edges of an antenna can also achieve improved gain and high directivity.

The proposed design of the UWB slotted antipodal bowtie antenna is etched on a 1.6 mm height FR4 substrate material whose relative permittivity is 4.4 and its dielectric loss tangent is 0.025. The antenna of dimension $30.4 \times 57mm^2$ is resonating at Ultra-Wideband frequency. The designing of a bowtie slot antenna starts with a radiating patch on the substrate, the ground plane as a reflector, and a Microstrip step feed line. Introducing two triangular-shaped mirror images at the upper and lower layers of the substrate imitates the antipodal structure in bowtie antenna. The current distribution of the antennas at the lower and middle band of UWB is optimized by incorporating the crossed slots. Placing a ground reflector beside the antenna improves its gain and directivity. Figure 13.1 [23] depicts the structure of the modified bowtie antenna with antipodal configuration. The step fed Microstrip feed line is improving the matching at the antenna edge. Table 13.1 describes about the dimensions of the design of antipodal slotted bowtie antenna.

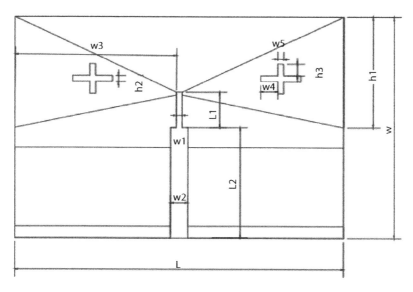

Figure 13.1 Structure of reflector-based antipodal bowtie antenna.

Table 13.1 Design parameters of antipodal slotted bowtie antenna [23].

Design label	Design parameters	Calculated values (mm)
W	FR4 Substrate width	37.5
L	FR4 substrate length	57
W1	The first step fed width	1
L1	Length of the first step fed	6
W2	Width of the second step fed	3
L2	Length of the second step fed	18.75
W3	Width of the wing bowtie	28.125
h1	Height of the wing bowtie	18.75
W4	Width of the cross slot 1	8
h2	Height of the cross slot 1	1
W5	Width of the cross slot 2	1
h3	Height of the cross slot 2	5

Gautam *et al.* (2013) [3], have proposed the design of a bowtie antenna by inserting an L strip for reducing the size of an antenna. Hassaine *et al.* (2012) [4], used the design of a slotted bowtie antenna for improving the bandwidth. Since the bowtie antenna is suitable for UWB applications by its multiband performance. The curved slot in the design of an antenna improves the impedance matching to make it perform for the UWB applications. Choi *et al.* (2004) [5], delineated the UWB antenna is designed using a single slot in the patch which is suitable for UWB applications. The UWB impedance bandwidth is maintained with a stable gain and radiation pattern. Kush Agarwal *et al.* (2011) [6], examined that the directional type of an antenna is framed by the design of a reflector at the bottom of the substrate. This design has an impact on the improvement of impedance bandwidth, axial ratio, and better VSWR results.

The general method of adjusting the cross slot width, length, and position helps to attain the required resonance frequency.

13.3.2   Fabrication of Slotted Bowtie Antenna with Reflector Prototype

Figure 13.2 (a) & (b) [23] shows the fabricated antenna on FR4 substrate with a dielectric constant of 4.6 and thickness of 1.6 mm. The RF–SMA (Radio Frequency- Surface Mount Adaptor) is used to exist the MPA. It consists of five-pin connectors. A monopole antenna with defective ground structure results in UWB and obtains a unidirectional pattern [20].

The fabricated bowtie antenna with the slot is shown in Figure 13.2 [23]; the return loss is measured using Agilent VNA via the RF-SMA port.

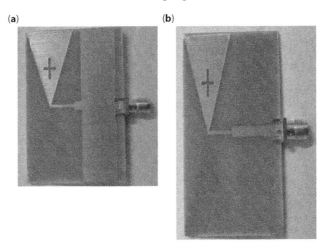

Figure 13.2 Fabricated slotted bowtie antenna with reflector: (a) top region and (b) bottom region.

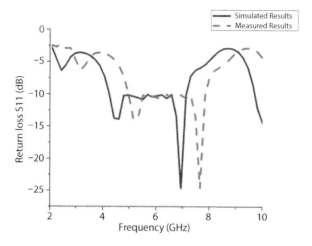

Figure 13.3 Simulated and measured return loss.

The various resonating frequencies like such as 4.6GHz, 6.94GHz which radiates in the UWB range with better return loss below -10dB. The bandwidth covered in this frequency range is of 2.48GHz. The simulated and VNA measured results of return loss are depicted in Figure 13.3 [23].

13.3.3   Parametric Study on the Effect of Slot in the Bowtie Antenna

Abbosh *et al.* (2006) [7], have designed the tapered slot planar antenna for the microwave imaging system. The slot is included in the patch antenna and improves the impedance bandwidth for UWB frequency range and gain of an antenna. Chen *et al.* (2000) [8], have examined the broad bandwidth with proper impedance matching can be achieved by the design of planar monopole antennas with slits at the radiator. In the slotted antipodal bowtie antenna, cross slots are inserted in the middle part of the antenna, which improves the impedance of an antenna. The width of the cross slot varies from 0.5mm to 1.5mm with a 0.5mm difference. This optimization in the width of slot 2 results in better return loss at -25dB for 6.94GHz at $W_5 = 1$mm. In the same manner, the height of slot 2 varies between 4mm to 6mm. The optimization in the height of the slot results in better return loss at -24dB for 6.94GHz at $h_3 = 5$mm. The width and height of slot1 are also optimized and better return loss is achieved at h2 = 1mm and W4 = 8mm. Better impedance matching is optimized by varying the length of the step feed 1 and 2 at $L_1 = 6$mm, $L_2 = 18.75$mm. The results of the effect in width, height of the slot, and length of the feed are shown in Figure 13.4-13.7 [23].

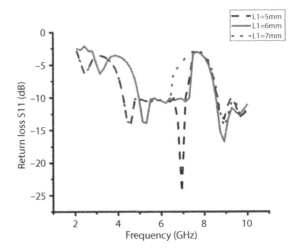

Figure 13.4 Simulated antipodal bowtie antenna return loss of effect in length of the first step fed L1 on slotted patch.

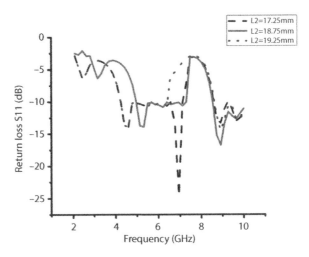

Figure 13.5 Simulated antipodal bowtie antenna return loss of effect in length the second step fed L2 on slotted patch.

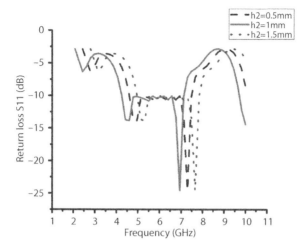

Figure 13.6 Simulated antipodal bowtie antenna return loss of effect in height of the cross slot 1 h2 on slotted patch.

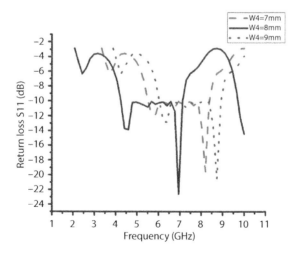

Figure 13.7 Simulated return loss of effect in the width of the cross slot 1 W4 on the slotted antipodal bowtie antenna.

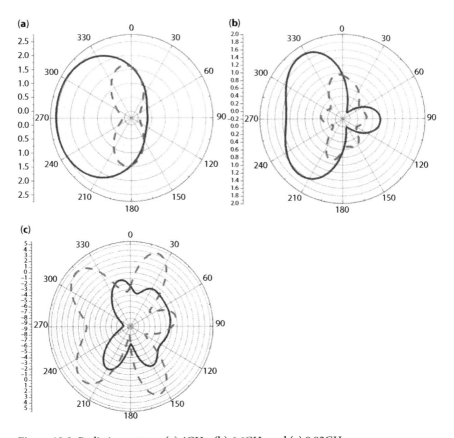

Figure 13.8 Radiation pattern: (a) 4GHz, (b) 6.6GHz and (c) 8.92GHz.

13.3.4   Radiation Pattern

The measurement of the radiation pattern of the antenna is carried out in simulated results. The radiation of the bowtie antenna is shown in Figure 13.8 [23], Comparison of the radiation pattern of measured and simulated are more satisfied at 4GHz, 6.6GHz, and 8.92GHz. Minimum back lobes are observed in the E-plane patterns and are directional. The angle of radiation is shifted from -50° to -130° over the frequency variation from 3.5GHz to 9GHz. A bi-directional ration is observed in the H-plane.

13.4   Design of a UWB-Based Compact Rectangular Antenna

Initially, a rectangular Microstrip patch antenna is designed for a centre frequency of 5.8 GHz. Similarly, FR4 substrate whose dielectric constant

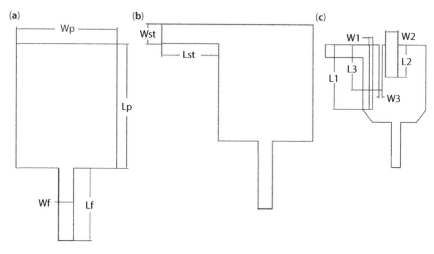

Figure 13.9 Evolution of the Rectangular MPA: (a) Rectangular MPA, (b) MPA with attached strip and (c) MPA with various slots.

of 4.4 and thickness h = 1.588 mm is considered for determining the dimension of the patch antenna. The development of the proposed Microstrip patch antenna is given in Figure 13.9 [23]. The calculated rectangular patch dimensions resonate at 5.8 GHz and obtain a minimum $|S_{11}|$ of -10 dB over a sharp bandwidth of 4.5 GHz to 6.5 GHz. Chen *et al.* (2007) [9], have proposed a compact UWB antenna with excellent radiation performance. The proposed UWB antenna geometry is shown in Figure 13.9 [21], and the optimal dimensions of the proposed UWB antenna are shown in Table 13.2. The layout of the proposed design in shown in Figure 13.10 [21].

13.4.1   Parametric Results of the Strip Attached at the Top of the Antenna

Miniaturization of antenna aperture makes the patch resonate mostly at high frequency because the wavelength of the metal radiator is inversely proportional to the frequency of resonance. As a result of the miniaturization, the fringing fields are bounded within the substrate, which increases the surface wave effects [12]. The gain, antenna efficiency, and narrow bandwidth are obtained from the sudden improvement in surface wave effects. Introducing any defect in the ground plane improves the fringing field between the patch and the bottom ground plane, which in turn decreases the surface wave effects. After the implementation of the

Table 13.2 Optimal dimensions of the miniaturized UWB antenna design.

Antenna parameters	Antenna dimensions (in mm)
Strip Length (L_{st})	6
Strip Width (W_{st})	2
First Slot Length (L_1)	10
First Slot Width (W_1)	0.5
Second Slot Length (L_2)	5
Second Slot Width (W_2)	2
Third Slot Length (L_3)	7
Third Slot Width (W_3)	0.5
Gap (g)	3.5
Length of the FR4 substrate (L_s)	22
Width of the FR4 substrate (W_s)	17
Triangle Slit Length (L)	2
Triangle Slit Width (W)	1.5

DGS, improvement occurs in operating bandwidth at high frequency from 5.8 GHz to 10.5 GHz. Generally, the low-frequency microwave signals only propagate into the human body whereas the high-frequency signals reradiate from the outer layer of the human body. Hence the increase in the length of the antenna results in a low-frequency resonance. Padmavathy & Madhan (2015) [13] have proposed a UWB patch antenna with bandwidth enhancement with multiple slots on a finite ground plane.

Vieira *et al.* (2017) [14] have studied planar inverted-F antenna wireless mobile applications. The Specific Absorption Rate (SAR) of the planar antenna on the human head model is detailed, which implies the signal penetration at the lower frequency range. Liu *et al.* (2017) [15] have proposed the performance and design principle of the planar inverted-F antenna. From the concept of planar inverted-F and inverted-L strip antenna for wireless communication, the proposed antenna is designed.

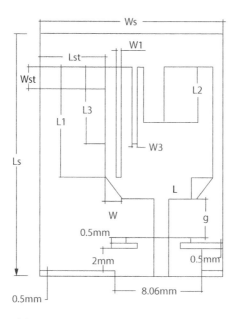

Figure 13.10 Layout of the miniaturized ultra-wideband antenna with DGS.

Attaching a small strip of length L_{st} and width W_{st} to the top of the patch antenna increase the patch length is shown in Figure 13.9 (b) [21]. The strip attached initiates the current flow distance from the feed input to the top end of the patch strip. Since the patch length is the fundamental parameter of antenna design, the increase in length gradually decreases the resonating frequency of the antenna. Monopole antenna with step edges improves the antenna bandwidth with multiple resonances [22].

The miniaturized UWB antenna structure enlarges the patch length which makes the antenna resonate in the lowest frequency spectrum. It minimizes the peak that arises in the reflection factor at 4 GHz shown in Figure 13.11 [21]. Further increase in the wavelength of the current flow is obtained by offsetting antenna feed on the right side of the antenna. The improvement in the length of the current flow is because of the 6.5 mm horizontal strip length at the top of the patch, a 10 mm vertical patch length, and a 4.5 mm horizontal patch length up to the centre of the feed strip. This structure of the attached strip with the antenna depicts an inverted-L shaped strip of length 21 mm. Hence taken as a whole the dimension of the antenna looks smaller and results in a minimum reflection coefficient of $|S_{11}| < -10$ dB at 3.4 GHz. In modern portable mobile devices, the most

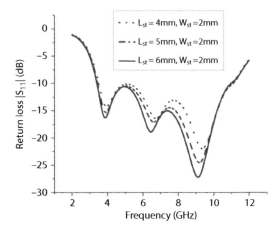

Figure 13.11 Effect of varying the strip length L_{st}.

popular methods of decreasing antenna dimensions are inverted-L and planar-F antennas.

13.4.2   Effect of Inserting Slot $L_1 x W_1$ and Location of the Slot d_s

Generally, the low-frequency signals penetrate deeper into the human tissues. It is observed that the reflection factor of the antenna is better only at higher frequencies as seen in Figure 13.12 [21]. Hence implementing a slot on the patch radiator varies the impedance bandwidth of the antenna. Usually, the length of the slot is half the guided wavelength is given in Equation (13.1).

$$L_{slot} = 0.45\lambda_g \qquad (13.1)$$

where λ_g is the guided wavelength of the slot.

In addition to the implementation of the slot of length L_1 and width W_1 on the patch radiator, the location of the slot d_s varies the antenna resonance characteristics as shown in Figure 13.12. Choosing a proper dimension and the location of the slot optimizes the operating frequency of the radiator. Not much variation can be seen in optimizing the slot width W_1. The variation in the position of the slot from the patch edge d_s optimizes the width of the inverted L shape on the patch. It correspondingly varies the resonance frequency and increases the slot length which shifts in the antenna resonance frequency towards lower frequency shown in Figure 13.12.

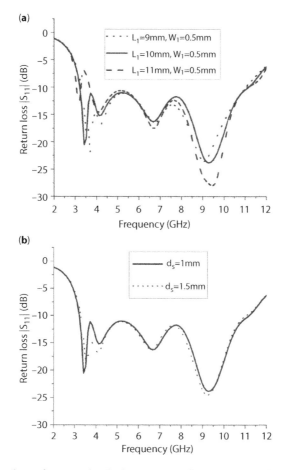

Figure 13.12 Effects of varying: (a) slot length L_1 and (b) slot position d_s.

13.4.3   Effect of Varying the Length of Slot L_2 and L_3

In the near future, the design of compact antennas with wideband characteristics is a significant research challenge for antenna engineers.

The improvement in impedance matching at the mid-band frequency is accomplished by etching different slots on the patch. Decreasing the patch length introduces changes mostly at the highest frequencies. In the proposed miniaturized antenna design, the minimum reflection coefficient $|S_{11}|$ less than -10 dB is accomplished towards the frequency range of 3.1 to 10.6 GHz on introducing various slots on the top layer of the antenna. The length of the second slot is L_2 and the third slot is L_3 are etched on the

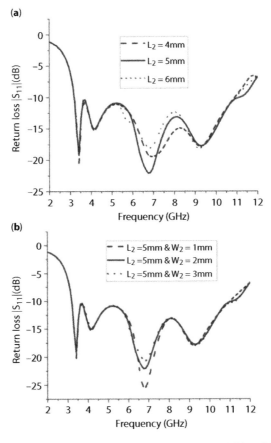

Figure 13.13 Effects of varying the second slot: (a) length L_2 and (b) width W_2 for a constant length.

patch, which results in the bandwidth improvement around 6.5 to 10 GHz as shown in Figure 13.13(a) and (b) [21].

13.4.4   Performance Comparison of the Measured and Simulated Results of the Miniaturized UWB Antenna

The proposed compact UWB antenna is fabricated on the FR4 substrate as shown in Figure 13.14 (b) [21]. The dimensions of the miniaturized UWB patch antenna are 17×22×1.58 mm^3 along with the Microstrip feed line technique. They are optimized to obtain proper impedance matching

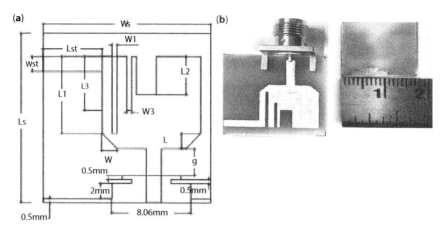

Figure 13.14 Proposed miniaturized UWB patch antenna: (a) Antenna layout and (b) Top and bottom of the fabricated antenna.

between the feed line and antenna edge to operate in the required frequency. The presence of the multiple slots on the patch antenna suppresses the ground plane effects.

The fabricated antenna is tested with the Field Fox Handheld VNA. Figure 13.15 [21] depicts that resonant frequency shift is observed between the VNA measurement of the fabricated antenna and the simulation results are because of soldering the Sub Miniature version A (SMA) connector

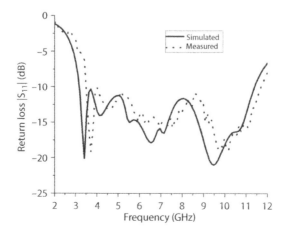

Figure 13.15 Comparison of simulated and measured results of frequency vs. return loss $|S_{11}|$.

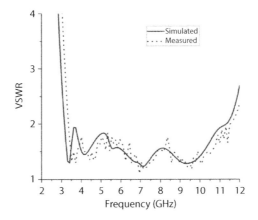

Figure 13.16 Comparison of simulated and measured results of frequency vs. VSWR [21].

with the antenna feed line. The soldering introduces some mismatch as at the input port. Figure 13.16 depicts the comparison of simulated and measured results of miniaturized UWB antenna.

13.4.5   Radiation Characteristic of the Proposed Miniaturized UWB Antenna

The E plane and H plane radiation patterns for various resonant frequencies of 3.4 GHz, 6.5 GHz, and 9.5 GHz of the antenna are shown in Figure 13.17 [21]. A resultant gain of 2 dBi is achieved from the compact microstrip patch antenna. The simulated antenna pattern represents that the antenna is omnidirectional at low frequency and becomes directional at a higher frequency.

13.5   Validation of the Miniaturized UWB Antenna with the Human Breast Model Developed

The designed miniaturized UWB patch antenna is arranged in a circular array to examine the tumour location in the phantom developed. For this, a 1 × 4 antenna array system arrangement is considered. The single transmitter and receiver antenna setup is used to study the dielectric strengths of the low and high coupling medium. Then the proposed compact UWB

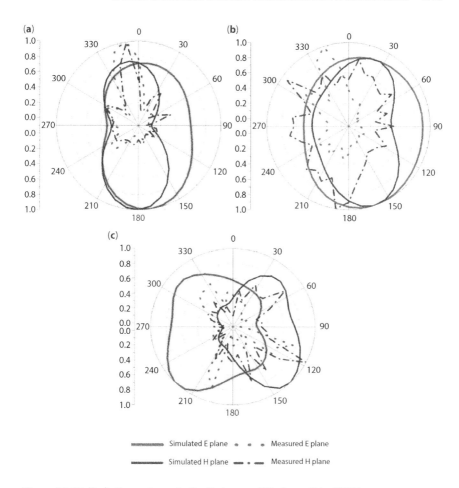

Figure 13.17 Radiation patterns in the E plane and H plane of the UWB antenna at frequencies: (a) 3.4 GHz, (b) 6.5 GHz, and (c) 9.5 GHz.

antenna is immersed in the coupling medium surrounding the human breast model. The typical scattering parameters are measured using the VNA connected to the transmitter and receiver antennas surrounding the breast phantom. The various stages of the tumour sizes are made from the proper mixture of wheat plus water ratio. These stages and types of tumour are used to define malignant or benign tissues. In this, the growing and major stages of the tumour sizes are considered for examination.

13.5.1   Validation of the Staircase UWB Antenna with the Human Breast Model Developed

The equivalent dielectric human breast model proposed by Lazebnik *et al.* (2007) [15] is employed for various researches on microwave imaging. The human breast model developed is the same as the Debye testbed setup of the compressed breast tissue-equivalent materials. The proposed 1×4 antenna array arrangement located on opposite sides to the human breast model for validation is shown in Figure 13.18 [21]. The single transmitter and receiver antenna setup is utilized to study the dielectric strengths of the low and high coupling medium. Then the proposed Staircase UWB antenna is immersed in the coupling medium surrounding the human breast model. Further, the typical scattering parameters are measured using the VNA connected to the transmitter and receiver antennas surrounding the breast phantom.

The American Society of Clinical Oncology (ASCO 2018) has approved the classification of the growth of cancer tissue in stages. The doctors define the cancer tissue from stage 0 to stage 4 which is used to define the size of the tumour (measured in centimetres (cm)), and its location. The tumour size considered from 1 cm to 5 cm falls under the specific stage and defines the growth of the tumour from breast tissue into the chest wall and skin. In the present research, four stages of tumour form size 1 cm to 4 cm are considered for validation. The proposed tumour size is further categorized as a malignant and benign tumour. The malignant tumour is more viscous due to high water concentration; hence it is considered as Stage III (3 cm) and Stage IV (4 cm), they are the growing and major

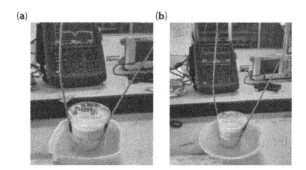

Figure 13.18 Proposed UWB patch antenna with Debye testbed model: (a) Lower dielectric and (b) Higher dielectric coupling medium.

stages of tumour respectively. The remaining Stage I (1 cm) and Stage II (2 cm) are the earlier and minor stages of cancer tissues, respectively. The wheat plus water mixture is considered as the tumour since it matches the assumption of tumour stages. The wheat plus water ratio defines the stages and type of tumour that might be malignant or benign.

13.5.2   MUSIC Beamforming Algorithm

The signal received from the scanning environment is collected by the array system to improve the signal strength along the desired direction of propagation. This methodology of beamforming can be adapted both at the transmitting and at the receiving array system to accomplish spatial selectivity. In recent days, it is used in various applications like sonar, radar, as a diagnostic device in medical imaging and wireless communication. In the transmitter antenna array system, the beam former controls the phase and amplitude of the single radiator. The beam former algorithm creates a constructive and destructive interference pattern in the wave front. While acting as receiver, the signals received from different directions are combined in such a way to detect the Direction of Arrival (DOA) of the received signals.

The antenna radiates low energy signals into the human breast model which is under the screening environment. If there is a presence of a tumour inside the human breast model, the intensity of the scattering effect is different from the signals reflected from the healthy tissue. Hence the amplitude of the signal varies according to the tumour dimensions. The desired signal is the signal scattered from the tumour. The received signal at the antenna terminals is superimposed with both desired and undesired (noise) signals. The undesired signals considered areas the noise interference from the healthy tissue. The magnitude of the received signals gives clear information about the tumour response. The Power Spectral Density (PSD) of the received signals is estimated to differentiate the tumour from the healthy tissues. Multiple Signal Classification (MUSIC) is proposed by Schmidt (1986). This MUSIC algorithm is the most accepted algorithm among the DOA estimation theory. Considered an M-array, vector elements are arranged linearly. MUSIC algorithm determines the noise subspace of the correlation matrix generated from the M-array antenna steering vector arrangement.

The direction vectors of the incident signals from the target are named steering vectors. The ideal response of the array signal source is called steering vectors, and it is orthogonal to the noise signal subspace.

Belhoul *et al.* (2003) [16] have constructed a real-life model to analyze the algorithm. At first, start with the general assumption of many plane waves from M narrowband sources. The plane waves generated from different angles θ_i = 1, 2, 3.....M incident on the N-equispaced sensors. Any time instant t, where t=1, 2...K, and K is the total number for snapshots. The output of the array system is the signal vector that consists of signal plus noise information (Shubair *et al.*, 2007) [17].

The signal vector x (t) is represented in Equation (13.2).

$$x(t) = \sum_{m=1}^{M} a(\theta_m) \cdot s_m(t) \qquad (13.2)$$

Where s (t) = Mx1 vector of generated waveforms, $a(\theta)$ = Nx1

$a(\theta)$ is the vector representation of the array antenna response to the source or array steering vector in that particular direction.

The $a(\theta)$ is represented in Equation (13.3).

$$a(\theta) = [1 \ e^{-j\varphi} \ \ldots\ldots\ldots\ldots e^{-j(N-1)\phi}]^T \qquad (13.3)$$

where T = transposition operator
   φ = electrical phase shift between elements in an array system

This electrical phase shift between each element is given in Equation (13.4).

$$\varphi = \frac{2\pi}{\lambda} d\cos\theta \qquad (13.4)$$

Where the element spacing is represented by variable d, this is usually assumed as half the wavelength of the highest frequency.

Additionally, the signal vector x(t) of size Nx1 is represented in Equation (13.5)

$$x(t) = A \, {}^{*}s(t) \qquad (13.5)$$

Where, the variable A is the steering vector and represented as A=[a(θ_1)....... a(θ_M)].

The output of the antenna array system consists of signal plus noise components and given in Equation (13.6)

$$y(t)= w(t)+x(t) \tag{13.6}$$

Where the assumed uncorrelated functions are, $y(t)$ is the output, and $w(t)$ is the noise. The function w (t) is considered as white noise with zero mean Gaussian processes.

Substituting Equation (13.5) in (13.6), in equation (13.6), the matrix size is $N \times K$ and represented in Equation (13.7)

$$U = A \cdot S + W \tag{13.7}$$

Where source waveforms $S = [s(1).......s(K)]$ of matrix size $M \times K$ and the noise vector $W = [w(1) .... w(K)]$ with $N \times K$ matrix.

The observed signal vector is represented by spatial correlation matrix R, is represented in expression (13.8)

$$R = E[u(t)u(t)^H] \tag{13.8}$$

Where the E[ ] – expectation operators and
H - the conjugate transpose operator

Substituting Equations (13.8) in (13.9), the spatial correlation matrix R is represented in Equation (13.9)

$$R = E[A \cdot s(t)s(t)^H \cdot A^H] + E[w(t)w(t)^H] \tag{13.9}$$

Assuming a signal model, whose correlation matrix R consists of signal eigenvalues M, and noise eigenvalues (N-M). The M eigenvalues are denoted as the matrix of $Es = [e_1 \ e_2 \ ......e_M]$, and the noise eigenvectors (N-M) are denoted as $En = [e_{M+1} \ e_{M+2}....e_N]$. In the power spectrum graph, whenever the steering vector $E (\varphi)$ is orthogonal to the noise subspace, the peaks are observed in the angular spectrum of the MUSIC algorithm. The equation of angular spectrum is given in Equation (13.10).

$$P_M(\varphi) = \frac{1}{E(\varphi)^H \cdot E_n \cdot E_n^H \cdot E(\varphi)} \tag{13.10}$$

13.5.3   Estimation of DOA Using MUSIC Algorithm

In the field of signal processing, the estimation of DOA provides the location of the signal source. The DOA provides the angle at which the signals have an incident on the antenna array [18]. Vaseghi (2008) [19] deals with various array signal processing algorithms applicable for audio, speech, and wireless communication. Knowing the angle of arrival of the signal sources incident on the antenna provides accurate identification of targets. The separation of the desired signals from the noise interferences invokes a beam steering at the required angle of the target location. The null-steering beam is invoked at the remaining angles of noise interferences. Hence from various DOA estimation algorithms, the MUSIC algorithm provides an accurate estimation of target location with high-resolution target identification.

Let the DOA estimation of the scattered signals from different sources. A circular array size of 1×4 antenna arrangement in a linear array surrounding the human breast model is considered for the validation process. Considering the signal sources at different angles, one is from the azimuth angle of $\varphi = 20°$ and an elevation angle of $\theta = 15°$. Another signal arrives from the azimuth angle of $\varphi = 30°$ and an elevation angle of $\theta = 25°$. The antenna elements are arranged in half of the wavelength concerning the highest resonance frequency. The observed DOA, the wave vector is determined. Then the matrix of the array response vector is evaluated. Additive Gaussian noise is considered as distortion in the signal processing

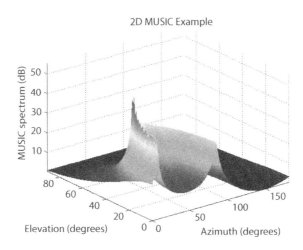

Figure 13.19 2D power spectrum plot using MUSIC algorithm for the signals received from the target.

environment. Eigenvalue decomposition is carried out in the received signal to create the eigenvector for the received signals. Finally, the equation of power spectrum estimation is applied to plot the 2D power spectrum plot of the signal received at the azimuth and elevation angles shown in Figure 13.19. The maximum peak point of 28 dB defines the resolution of the tumour location and the remaining minimum peak points are because of the interference.

13.6   Conclusions

The present research contributes to the development of more efficient Microstrip patch antennas for detecting breast tumours at the earliest stages making use of microwave imaging. In recent decades breast cancer is a major disease found among young women. Various screening techniques like X-ray imaging, Positron Emission Tomography (PET), and Magnetic Resonance Imaging (MRI) scanning depict the major role of medical devices in studying Breast tissue abnormalities. The existing screening techniques result in several misjudgements of tumours present in the breast tissue. Hence an efficient microwave imaging technique is required to examine breast cancer efficiently at the earliest growing stage.

The first part of the design and investigation of the slotted antipodal bowtie antenna with reflector is done for the dimension of $37.5 \times 57 \times 1.6$ mm^3. Slotted antipodal bowtie antenna exhibit a gain of 4.33dB and the directivity of 4.9dB with Fractional Bandwidth (FBW) of 50%. The improvement in FBW is accomplished towards the structure of the antipodal bowtie staircase antenna with a reflector of 57.48%. The antipodal bowtie staircase antenna is designed for the dimension of $30.4 \times 57 \times 1.6$ mm^3 and obtained the gain of 4.25dB with the directivity of 4.83dBi. The second part of a miniaturized Microstrip patch antenna with an inverted-L strip attached to the top of the patch. The antenna patch length depends on the operating frequency of resonance; hence the decrease in antenna size makes the antenna resonate at higher frequencies. Its dimension is 17mm × 22mm × 1.588 mm. Also, the decrease in the radiating surface of the patch radiator increases the surface wave currents. This surface wave effect decreases the antenna efficiency, gain, and impedance bandwidth. The increase in the length of the strip and offsetting the feed line results in a longer current flow. This makes the antenna resonate at the lower edge of the UWB frequency of 3.4 GHz. A 9 GHz impedance bandwidth is accomplished by implementing the Defective Ground Structure (DGS) in the ground plane. This results in the antenna having a return loss of below -10 dB from 2.9 GHz to 11.9 GHz. The fractional bandwidth of the proposed antenna is 122% with an optimum gain of 2 dBi. The fabricated patch antenna is tested for scattering parameter and VSWR using VNA and compared with the simulated results. Finally, the depth of tumour location is estimated on an average of 44mm from the skin boundary. The accuracy of tumour depth identification is 88%. The DOA of the scattered signals is used to identify the exact location of the tumour from its power spectrum using the MUSIC algorithm.

References

1. The American Cancer Society Medical and Editorial Content Team. (December 8, 2015). What is cancer. http://www.cancer.org/cancer/cancer basics/what-is-cancer. Accessed Aug. 2019.
2. American Society of Clinical Oncology. ASCO 2018. Available from: https://www.cancer.net/cancer-types/breast-cancer/stages.
3. Gautam, AK, Yadav, S & Kanaujia, BK 2013, A CPW-fed compact UWB microstrip antenna, *IEEE Antennas and Wireless Propagation Letters*, vol. 12, pp. 151-154.
4. Seladji-Hassaine, N., Merad, L., Meriah, S.M. and Bendimerad, F.T., 2012. UWB Bowtie Slot Antenna for Breast Cancer Detection. *World Academy of Science, Engineering and Technology*, 71, pp. 1218-1221.
5. Choi, SH, Park, JK, Kim, SK & Park, JY 2004, A new ultra-wideband antenna for UWB applications, *Microwave and optical technology letters*, vol. 40, no. 5, pp. 399-401.
6. Agarwal, K & Alphones, A 2012, Wideband circularly polarized AMC reflector backed aperture antenna, *IEEE Transactions on antennas and propagation*, vol. 61, no. 3, pp. 1456-1461.
7. Abbosh, AM, Kan, HK & Bialkowski, ME 2006, Compact ultra-wideband planar tapered slot antenna for use in a microwave imaging system, *Microwave and optical technology letters*, vol. 48, no. 11, pp. 2212-2216.
8. Chen, ZN & Chia, YWM 2000, Impedance characteristics of trapezoidal planar monopole antennas, *Microwave and Optical Technology Letters*, vol. 27, no. 2, pp. 120-122.
9. Chen, ZN, See, TS & Qing, X 2007, Small printed ultrawideband antenna with reduced ground plane effect, *IEEE Transactions on antennas and propagation*, vol. 55, no. 2, pp. 383-388.
10. Barrett, TW 2012, Development of ultrawideband communications systems and radar systems, in JD Taylor, *Ultrawideband Radar*, 2017, CRC Press.
11. Ghavami, M, Michael, LB & Kohno, R 2007, *Ultra Wideband Signals and Systems In Communication Engineering*, John Wiley & Sons, Ltd.

12. Hamad, EK & Radwan, AH 2013, Compact ultra wideband microstrip-fed printed monopole antenna, In *30th National Radio Science Conference, Egypt.*

13. Padmavathy, AP & Madhan, MG 2015, An improved UWB patch antenna design using multiple notches and finite ground plane, *Journal of Microwaves, Optoelectronics and Electromagnetic Applications,*   vol. 14, no. 1, pp. 73-82.

14. Vieira, VF, Pessoa, LM & Carvalho, MI 2017, Evaluation of SAR induced by a Planar Inverted-F Antenna based on a Realistic Human Model, In *EMBEC & NBC*, pp. 599-602, Springer, Singapore.

15. Liu, NW, Zhu, L & Choi, WW, 2017, A low-profile wide-bandwidth planar inverted-F antenna under dual resonances: Principle and design approach, *IEEE Transactions on Antennas and Propagation*, vol. 65, no. 10, pp. 5019-5025.

16. Belhoul, FA, Shubair, RM & Ai-Mualla, ME, 2003, Modelling and performance analysis of DOA estimation in adaptive signal processing arrays, In *10th IEEE International Conference on Electronics, Circuits and Systems, 2003*. ICECS 2003. *Proceedings of the*, vol. 1, pp. 340-343.

17. Shubair, RM, Al-Qutayri, M & Samhan, JM, 2007, A Setup for the Evaluation of MUSIC and LMS Algorithms for a Smart Antenna System, *JCM*, vol. 2, no. 4, pp. 71-77.

18. Kim, H & Viberg, M 1996, Two decades of array signal processing research. *IEEE signal processing magazine*, vol. 13, no. 4, pp. 67-94.

19. Vaseghi, SV 2008, Advanced digital signal processing and noise reduction, John Wiley & Sons.

20. Vijayalakshmi, J & Murugesan, G,   A miniaturized high gain (MHG) Ultrawide band unidirectional monopole antenna for UWB applications, *Journal of Circuits, Systems, and Computers*, vol. 28, issue 14, pp. 1950230-1950249.

21. Dinesh V. and Govindasamy, M., 2019 A miniaturized planar antenna with defective ground structure for UWB applications, *IEICE Electronics Express*, 16(14), pp. 20190242-20190242.

22. Dinesh, V. and Murugesan, G., 2018, December, A Compact Stair Case Monopole UWB Antenna for radar Applications, In *2018 International Conference on Intelligent Computing and Communication for Smart World (I2C2SW)* (pp. 314-316). IEEE.

23. Vijayalakshmi, J. and Murugesan, G., 2018, December. Design of UWB High Gain Modified Bowtie Antenna for Radar Applications. In *2018 International Conference on Intelligent Computing and Communication for Smart World (I2C2SW)* (pp. 311-313). IEEE.

24. Bulyshev, AE, Souvorov, AE, Semenov, SY, Posukh, VG & Sizov, YE 2004, Three-dimensional vector microwave tomography: theory and computational experiments. *Inverse Problems*, vol. 20, no. 4, pp. 1239.

Joint Transmit and Receive MIMO Beamforming in Multiuser MIMO Communications

**Muhammad Moinuddin[1,2]*, Jawwad Ahmad[3], Muhammad Zubair[4]
and Syed Sajjad Hussain Rizvi[5]**

[1]*Center of Excellence in Intelligent Engineering Systems (CEIES),
King Abdulaziz University, Jeddah, Saudi Arabia*
[2]*Electrical and Computer Engineering Department, King Abdulaziz University,
Jeddah, Saudi Arabia*
[3]*Electrical Engineering Department, Usman Institute of Technology, Karachi, Pakistan*
[4]*Department of Computer Science, Iqra University, Karachi, Pakistan*
[5]*Computer Science Department, SZABIST, Karachi, Pakistan*

Abstract

Multiuser MIMO is one of the candidates for 5G technology as it promises high throughput and better spectral efficiency. Most of the existing work in Multiple Input Multiple Output (MIMO) beamforming employs the single-sided adaptation of beam (i.e., either transmit or receive). In addition to that, there a lot of work that has been presented using LMS-based algorithms, which is well known to perform better in the Gaussian environment. Inherently, the LMS-based algorithms have slow convergence. In this work, we propose a novel joint transmit and receive MIMO beamforming architecture that utilizes a decision-directed structure. It consists of two phases: a training phase in which the error of the algorithm is calculated via training symbols and a decision-directed phase where the error is generated from the estimated symbols at the output of the adaptive beam-former. To deal with the non-Gaussian environment, we propose to use a Generalized Least Mean (GLM) algorithm. Consequently, a recursive weight update is developed for both transmitting and receiving antennas. The performance of the proposed beamforming is investigated in the presence of frequency selective Rayleigh fading channel. Simulation results are presented to corroborate the theoretical findings.

**Corresponding author*: mmsansari@kau.edu.sa

Prashant Ranjan, Dharmendra Kumar Jhariya, Manoj Gupta, Krishna Kumar, and Pradeep Kumar (eds.)
Next-Generation Antennas: Advances and Challenges, (251–262) © 2021 Scrivener Publishing LLC

Keywords: MIMO communications, multiuser MIMO, beamforming, mean square error, convergence analysis, mean square stability, generalized least mean, joint optimization

14.1   Introduction

Wireless communication is a very fast-growing field these days. Presently wireless communication is exceptionally imperative for us, and thus we utilize it immensely [1, 2]. In recent years, MIMO adaptive antennas have been used as one of the most popular architectures to meet the higher throughput requirements. The MIMO system is the most advanced architecture for improving bandwidth efficiency and overall performance [3–6].

In the literature, several results have been reported for MIMO beamforming [1–9]. It gives substantial space-time communications against fading channels [10]. Pham *et al.* presented signal-to-noise and interference ratio (SNIR)–based design for both transmit and receive beamforming in MIMO system [10]. It assumes that channel state information is available. The same authors in [11] analyzed multiuser MIMO beamforming under frequency selective channels by employing a similar criterion of SNIR. To minimize mean-square-error (MSE) in the mobile environment, the efficiency of both Least Mean Square (LMS) and Normalized LMS (NLMS) is compared at the receiver's weight vectors [12]. In [13], a neural network–based approach is used only at the receiver end for beamforming and interference cancellation. The goal was to attain a Wiener solution. It was found that systems employing these functions are successful in tracking mobile users in real time. In [14], a class of linear derivative constraints is presented to handle broadband and moving jammer sources problems on the conventional narrowband uniform linear array configuration. The null width is obtained by robust modification of linear constrained minimum variance (LCMV) algorithm in the jammer directions. In [15], an alternative approach was provided to design transmit beamforming in a multiuser scenario that maximizes the signal-to-leakage ratio (SLR).

In this work, we propose a novel MIMO beamforming solution that employs generalized least mean (GLM) family of adaptive algorithms for obtaining the beamforming weights. The novelty of the proposed work resides in the fact that GLM adaptive algorithms have never been tested for MIMO beamforming. The rationale behind using GLM-based adaptive algorithms is that MIMO beamforming in the presence of fading channel experiences a non-Gaussian environment, and it is a well-established fact that LMF performs better in the non-Gaussian environment. Moreover, convergence analysis is carried in both the mean and the mean square sense. Consequently, the stability bounds that ensure both mean and mean-square-error stability are provided.

Following the introduction, the proposed model is presented. The next section is dedicated to the MIMO beamforming based on the generalized least mean algorithm. Before simulating the results, the mean and the mean-square-error stability of the adaptive algorithm is presented. Finally, concluding remarks are presented.

14.2 System Model: Proposed Mimo Beamforming Architecture

The proposed generic architecture of MIMO beamforming, as discussed in [16], is shown in Figure 14.1 with the modification of decision-directed mode. It is an $M \times N$ MIMO architecture with a single data stream over frequency selective fading channel. The working concept of the proposed model is that the transmit signals at one end are "combined" with the receive signals at another end. The combination must be in a way to achieve an improved bit error rate or data rate for each MIMO user. The number of maximum cancellation delayed channels L, also known as the degree of freedom, is computed in [17].

The output equation for the above-proposed model from the receiver (in the presence of channel state information (CSI)) is given as:

$$y(n) = \mathbf{w}_r^H(n)\, \mathbf{A}^0 \mathbf{w}_t(n)\, s(n)$$

$$+ \sum_{i=1}^{L} \mathbf{w}_r^H(n)\, \mathbf{A}^i \mathbf{w}_t(n)\, s(n - \tau_i) + \mathbf{w}_r^H(n)\, \mathbf{v}(n) \qquad (14.1)$$

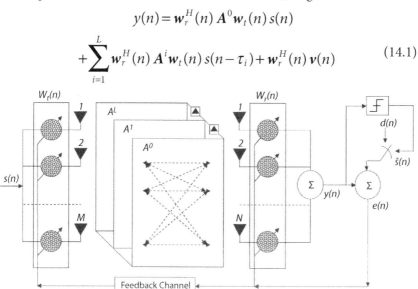

Figure 14.1 MIMO channel beamforming architecture with decision-directed mode [16].

Where $s(n)$ is the transmitted signal, τ_i is the delay for the ith multipath, $\mathbf{A}^0$ and $\mathbf{A}^i$ are, respectively, the direct and i^{th} delayed $N \times M$ channel matrix, $\mathbf{w}_t(n)$ is the $M \times 1$ transmit beamforming weight vector, and $\mathbf{w}_r(n)$ is the $N \times 1$ received beamforming weight vector. Here, $\mathbf{v}(n)$ is $N \times 1$ vectorshowing zero mean additive white Gaussian noise (AWGN) with variance δ_v^2.

For simplicity, it is assumed that every delayed input signal is uncorrelated with zero mean value, therefore

$$E[s^*(n - \tau_j)s(n - \tau_k)] = 0 \quad for\, j \neq k \tag{14.2}$$

14.3   Mimo Beamforming Based on Generalized Least Mean (GLM) Algorithm

In this chapter, we consider the GLM algorithm, whose cost function is given in [18] and is shown in Equation (14.3). The cost function shows $2K^{th}$ power of the least mean error. If $K = 1$, it will become an LMS algorithm which does not perform well in the non-Gaussian environment. Since the channel under consideration is Rayleigh, it is expected to have better performance of GLM that is $K > 1$.

$$J(n) = E[|e(n)|^{2K}] \tag{14.3}$$

Here $e(n)$ is the error for the Least Mean adaptive algorithm, which is calculated differently in the two phases of weight adaptation. In the training phase, the error is calculated via:

$$e(n) = s(n) - \mathbf{w}_r^H(n)\, \mathbf{u}(n) \tag{14.4}$$

While in the decision-directed phase, the error is estimated as:

$$e(n) = \hat{s}(n) - \mathbf{w}_r^H(n)\, \mathbf{u}(n) \tag{14.5}$$

where $\mathbf{u}(n)$ is given as:

$$\mathbf{u}(n) = \mathbf{A}^0 \mathbf{w}_t(n)\, s(n) + \sum_{i=1}^{L} \mathbf{A}^i \mathbf{w}_t(n)\, s(n - \tau_i) + \mathbf{v}(n) \tag{14.6}$$

14.3.1   Update of the Receive Weight Vector

By employing the steepest descent optimization, the generic receive weight update equation for GLM algorithm is written as:

$$w_r(n+1) = w_r(n) - \mu \nabla_{w_r} J(n) \tag{14.7}$$

where ∇_{w_r} represents the gradient operation with respect to receiving weight vector w_r. Thus, by evaluating $\nabla_{w_r} J(n)$ for GLM algorithm the final received weight update equation becomes

$$w_r(n+1) = w_r(n) + 2\mu Ke(n)^{2K-1} u(n) \tag{14.8}$$

14.3.2   Update of Transmit Weight Vector

Using a similar approach, as presented in section 14.3.1, the generic transmit weight update equation for GLM algorithm is given by:

$$w_t(n+1) = w_t(n)$$

$$+2\mu Ke(n)^{2K-1} \left[w_r^H(n) \left(A^0 s(n) + \sum_{i=1}^{L} A^i s(n-\tau_i) \right) \right]^H \tag{14.9}$$

14.4   Mean and Mean Square Stability of the GLM

In order to analyze the stability of the GLM, the correlation of the inputs to transmitting and receiving antennas needs to be evaluated. The correlation of the input to the transmitting antenna is given by $z(n)$ as follows:

$$R_z = E[z(n)z^T(n)] \tag{14.10}$$

where

$$z(n) = \left(A^{0^T} s^*(n) + \sum_{i=1}^{L} A^{i^T} s^*(n-\tau_i) \right) w_r(n)$$

The correlation of the receiving weights is given by an $N \times N$ matrix, C_{W_r},

$$R_z = A^{0^T} C_{W_r} A^0 \delta_s^2 + A^{i^T} C_{W_r} A^0 \delta_s^2 + A^{0^T} C_{W_r} A^i \delta_s^2 + A^{i^T} C_{W_r} A^i \delta_s^2$$

Using Equation (14.6), the correlation of the input to receiving antenna u is given by:

$$R_u = E[u(n)u^T(n)] \tag{14.11}$$

The correlation of the transmitter weights is given by an $M \times M$ matrix, C_{W_t},

$$R_u = A^0 C_{W_r} A^{0^T} \delta_s^2 + A^0 C_{W_r} A^{i^T} \delta_s^2 + A^i C_{W_t} A^{0^T} \delta_s^2 + A^i C_{W_t} A^{i^T} \delta_s^2 + \delta_v^2$$

Next, by defining the following correlation matrix

$$R \triangleq \begin{cases} R_u \text{ Correlation matrix for transmitter weights} \\ R_z \text{ Correlationmatrix for receiver weights} \end{cases}$$

and using the approach of [18], the mean and the mean square stability can be guaranteed by the following conditions:

$$0 < \mu < \frac{1}{K N (2K-1)\delta_v^{2K-2} R} \text{ for mean stability} \tag{14.12}$$

and

$$0 < \mu < \frac{\delta_v^{2K-2}}{K(4K-3)NR \, \delta_v^{4K-4}} \tag{14.13}$$

for mean square stability.

14.5   Simulation Results

For simulating the results, we have investigated frequency selective Rayleigh channel with different numbers of multipaths (L). The Rayleigh channel is simulated using Jakes Model. The transmitted data used is BPSK modulated.

In order to examine the performance of the proposed MIMO beam-former, effects of the following various parameters have been observed:

- Effect of K
- Effect of μ
- Effect of M and N
- Effect of SNR

In the ensuing, we have discussed all the above-mentioned parameters individually.

14.5.1   Effect of K on the MSE

Figure 14.2 shows the impact of values of K on the mean square error (MSE). Three different values of K have been considered while keeping other parameters constant. The value of SNR is considered as *20 dB*, both M and N are set to *3*, and the step size μ is fixed at *0.001*. The graph indicates improved MSE by increasing the value of K. The best steady state for MSE is achieved at $K = 10$.

14.5.2   Effect of μ on the MSE

The effects of the *step size (μ)* have been investigated and shown in Figure 14.3. The value of μ varies from *0.01* to *0.0001* for the same value of $K =$ *2*. The SNR and number of transmitter/receiver antennas are the same as discussed above.

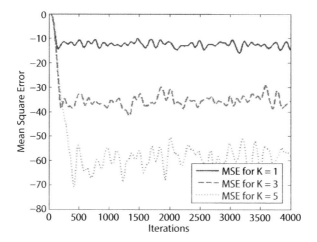

Figure 14.2 Mean square error with different values of K.

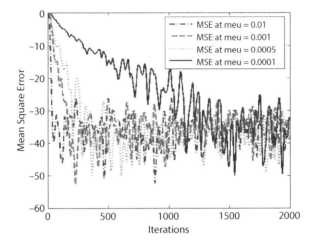

Figure 14.3 Mean-square-error for different values of μ.

14.5.3   Effect of *M* and *N* on the MSE

The effect of number of transmitter and receiver antennas (i.e., *M* and *N*) can be seen in Figure 14.4. Here the value of *K* is kept at 3 while SNR and step size values are *20 dB* and *0.001,* respectively. The results show that the increment in the number of transmitting and receiving antennas improve the MSE.

14.5.4   Effect of SNR on the MSE

The impact on MSE is investigated by varying the value of SNR and the results are given in Figure 14.5. However, the number of transmitter and

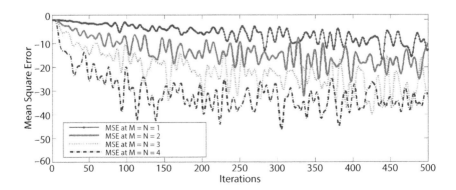

Figure 14.4 Mean square error with different number of transmitter and receiver antennas.

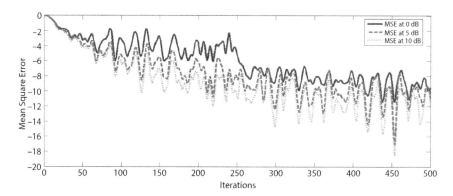

Figure 14.5 Mean square error versus SNR for the same iteration.

receiver antennas, the value of K and the step size μ are kept constant at *3, 4,* and *0.0001,* respectively. The graph clearly shows the improvement in MSE by increasing the value of SNR.

14.5.5   Effect of SNR on Bit Error Rate

Finally, the bit error rate versus SNR is evaluated on different numbers of transmitter and receiver antennas, shown in Figure 14.6. Again with this graph, the concept of diversity has been proved.

All the above results show that the weight update is very effective in the proposed architecture. Using this approach, bit error rate and mean square error are also improved.

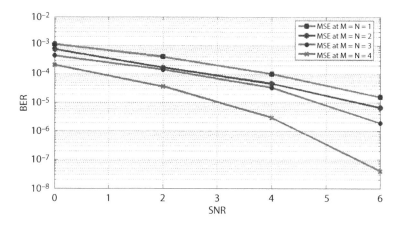

Figure 14.6 Decrease in bit error rate with the increase in number of transmitter/receiver.

14.6   Summary

We proposed a new MIMO beamforming architecture for joint transmit and receive antennas, which consist of a decision-directed structure. We also applied the GLM algorithm, which operates well in the non-Gaussian environment. The entire simulation is carried out in the presence of frequency selective fading channel. The proposed MIMO beamformer outperforms the conventional LMS algorithm by a significant margin. The mean square error performance is improved by increasing the values of K, the number of transmitter/receiver antennas, and the value of SNR. We also derived the necessary conditions for both the mean and mean square stability of the proposed algorithm in the context of MIMO beamforming.

References

1. Syed Sajjad Hussain Rizvi, Muhammad Zubair, Jawwad Ahmad, ManzoorHashmani, and Muhammad Waqar Khan. "Wireless Communication as a Reshaping Tool for Internet of Things (IoT) and Internet of Underwater Things (IoUT) Business in Pakistan: A Technical and Financial Review." *Wireless Personal Communications* (2019): 1-19.
2. Muhammad Zubair, Jawwad Ahmad, and Syed Sajjad Hussain Rizvi. "Miniaturization of Monopole Patch Antenna with Extended UWB Spectrum via Novel Hybrid Heuristic Approach." *Wireless Personal Communications* 109, no. 1 (2019): 539-562.
3. E. Telatar, "Capacity of Multi-Antenna Gaussian channels", *AT&T Bell Labs. Tech. Memo.*, vol.10, no.6, pp. 585–595, Oct. 1995.
4. K.J.R. Liu, F.R. Farrokhi, and L. Tassiulas, "Transmit and Receive Diversity and Equalization in Wireless Networks with fading channels", *Proc. IEEE Globecom*, pp. 1193–1198, Phoenix, AZ, Nov. 1997.
5. G.J. Foschini and M.J. Gans, "On limits of wireless communications in a fading environment when using Multiple Antennas", *Wirel. Pers. Communication*, vol.6, no.3, pp. 311–335, March 1998.
6. G.J. Foschini, G.D. Golden, R.A. Valenzuela, and P.W. Wolniansky, "Simplified processing for high spectral efficiency wireless communication employing multi-element arrays", *IEEE J. Sel. Areas Communication*, vol. 17, no.11, pp. 1841–1852, Nov. 1999.
7. M. A. Jensen and J. W. Wallace, "A Review of Antennas and Propagation for MIMO Wireless Communications", *IEEE Transactions on Antennas and Propagations*, vol. 52, no. 11, November 2004.

8. A. J. Paulraj, D. A. Gore, R. U. Nabar and H. Bolcskei, "An Overview of MIMO Communications—A Key to Gigabit Wireless", *Proceedings of the IEEE*, vol. 92, no. 2, February 2004.

9. Jawwad Ahmad, "Design of Efficient Adaptive Beamforming Algorithms for Novel MIMO Architectures". IQRA University (2014), Karachi.

10. H. H. Pham, T. Taniguchi, and Y. Karasawa, "The Weights Determination Scheme for MIMO Beamforming in Frequency-Selective Fading Channels", *IEICE Transaction on Communication*, vol. E87–B, no. 8, August 2004.

11. H. H. Pham, T. Taniguchi, and Y. Karasawa, "Multiuser MIMO Beamforming for Single Data Stream Transmission in Frequency-Selective Fading Channels", *IEICE Transaction on Communication*, vol. E88–A, no. 3, March 2005.

12. M. Yasin, P. Akhtar, and Valiuddin, "Performance Analysis of LMS and NLMS Algorithms for a Smart Antenna System", *International Journal of Computer Applications*, vol. 4, no. 9, August 2010.

13. A. H. El Zooghby, C. G. Christodoulou, and M. Georgiopoulos, "A neural-network-based Linearly constrained minimum variance beamformer", *Microwave and Optical Technology Letters*, vol. 21, no. 6, June 1999.

14. R. Li, X. Zhao, and X. W. Shi, "Derivative Constrained Robust LCMV Beamforming Algorithm", *Progress in Electromagnetics Research C*, Vol. 4, 43–52, 2008.

15. R.A.Tarighat, M. Sadek, and A. H. Sayed, "A Multi-User Beamforming Scheme for Downlink MIMO Channels Based on Maximizing Signal-To-Leakage Ratios", *IEEE International Conference on Acoustics, Speech, and Signal Processing*, pp. III 1129 – 1132, March 2005.

16. H. H. Pham, T. Taniguchi, and Y. Karasawa, "The Weights Determination Scheme for MIMO Beamforming in Frequency-Selective Fading Channels", *IEICE Transaction on Communication*, vol. E87–B, no.8, August 2004.

17. Shengli and G. B. Giannakis, "Space-time coding with maximum diversity over frequency-selective fading channels", *IEEE Signal Process. Letter*, vol. 8, no. 10, pp. 269–272, Oct. 2001.

18. E. Walach and B. Widrow, "The Least Mean Fourth (LMF) adaptive algorithm and Its Family", *IEEE Transactions on Information Theory*, vol. IT-30, no. 2, March 1984.

15

Adaptive Stochastic Gradient Equalizer Design for Multiuser MIMO System

Muhammad Moinuddin[1,2], Jawwad Ahmad[3], Muhammad Zubair[4]*
and Syed Sajjad Hussain Rizvi[5]

*[1]Center of Excellence in Intelligent Engineering Systems (CEIES),
King Abdulaziz University, Jeddah, Saudi Arabia
[2]Electrical and Computer Engineering Department, King Abdulaziz University,
Jeddah, Saudi Arabia
[3]Electrical Engineering Department, Usman Institute of Technology,
Karachi, Pakistan
[4]Department of Computer Science, Iqra University, Karachi, Pakistan
[5]Computer Science Department, SZABIST, Karachi, Pakistan*

Abstract

The demand for a very high data rate required for various upcoming applications in future 5G generation, such as machine-to-machine communications, video streaming, and Internet of Things (IoT), etc., is not possible to meet with the conventional multiple-input multiple-output (MIMO) system. To overcome these challenges, promising candidates for the 5G are Multiuser MIMO (MU-MIMO) and Massive MIMO (M-MIMO) as they promise to provide both higher throughput and broader coverage. However, these systems require efficient design of precoder and equalizer to improve the overall system performance. Typically, precoders and equalizers are designed via supervised techniques by estimation of the Channel State Information (CSI) at the base station using pilot sequences. However, this increases the overhead of transmission and hence degrades the spectral efficiency. To deal with these issues, we propose in this work to design a semi-blind adaptive equalizer. This will be based on minimizing the bit error rate (BER) expression via a stochastic gradient approach. For this, we first derive a closed-form expression for the BER in the MU-MIMO systems. This is then utilized to develop an iterative adaptive stochastic gradient equalizer. Simulation results are presented to investigate the performance of the proposed equalizer.

*Corresponding author: mmsansari@kau.edu.sa

Prashant Ranjan, Dharmendra Kumar Jhariya, Manoj Gupta, Krishna Kumar, and Pradeep Kumar (eds.)
Next-Generation Antennas: Advances and Challenges, (263–276) © 2021 Scrivener Publishing LLC

Keywords: 5G, MIMO systems, wireless communication, adaptive equalizer

15.1   Introduction

The multiple-input-multiple-output (MIMO) system refers to a communication scenario where both transmitter and receiver are equipped with multiple antennas. The MIMO system received great attention as it has proved to provide better spectral efficiency and higher data rate [1, 2]. In addition to these, it has been found that the channel capacity can be enhanced by increasing the number of transmit and receive antennas, that is, spatial multiplexing. In earlier proposals, the MIMO system was proposed to implement without considering co-channel users. However, when the MIMO system is designed for multiuser scenario, also known as multiuser MIMO (MU-MIMO) system, the co-channel users signal interferes with the desired signal which is termed as co-channel interference [3-6]. Another concept termed as multiuser diversity is also utilized to improve spectral efficiency in MU-MIMO systems [7].

The MU-MIMO in wireless communication brings advancement on four major fronts. These are [8]:

- Increase in data rate due to more number of antennas which allows more independent streams of data to be transmitted and serves more receiver terminals simultaneously.
- Enhanced reliability due to more number of antennas that provide many distinct paths through which the radio signal can be propagated.
- Improvement in energy efficiency due to directional antennas at the base station (BS).
- Reduction in interference as the BS is transmitting in the direction where spreading interference is harmless.

However, it is difficult to achieve ideal improvements in MU-MIMO due to various practical limitations. In addition to this, there exists a tradeoff in terms of system performance and complexity in implementation.

15.2   Related Literature Review

In the literature, there is a lot of work related to precoder design. The zero forcing (ZF) and the minimum-mean-square-error (MMSE) precoding techniques have shown better Multiple User Interference (MUI) cancelation performance [3, 9]. However, this is achieved at the cost of a decrease in data rate.

However, truncated polynomial expansion is used in massive MIMO systems rather than regularized zero-forcing (RZF) due to its complex implementation as it requires fast inversion for large matrices [10]. In [11], it is shown that by employing appropriate constraints a sub-optimal solution that maximizes the sum capacity can be achieved. Another technique in [12] utilizes an improved MMSE method for interference cancellation in MIMO system. A successive MMSE techniques was introduced in [13] which has shown better performance in comparison to the existing state-of-the-art techniques. Also, it has less computational complexity. In a recent approach, the implementation of Massive MU-MIMO is considered promising by incorporating a hybrid block diagonal (Hy-BD) [14]. In [15], the authors proposed a selective vector precoding scheme to minimize power using a perturbation method which was later improved for a large system in [16].

Equalization for MU-MIMO is a well-established research area with numerous existing works [17]. In this context, there are several techniques which have different compromises between the complexity of the algorithm and the performance of the overall system [18]. In [19], a variant of MMSE method was proposed that uses zero padding and it has been shown to achieve optimal diversity. In the same direction, the performance of decision feedback and linear MMSE equalizers are compared in [20] which was later improved by proposing a bidirectional decision feedback equalizer [21]. For MIMO-OFDM, an iterative equalizer in frequency domain was implemented in [22].

In [23], Minimum-BER (MBER) criterion was introduced to design equalizer which was later used to develop MBER-based decision feedback equalizer in [24] and [25]. However, the complexity of these algorithms is huge compared to the LMS type algorithms.

15.3   System Model

We consider a multiuser uplink scenario in which the mobile station (MS) comprises the desired user's in addition to K number of interfering user's that utilize a single transmitting antenna, as shown in Figure 15.1. The base station (BS) consists of M number of receiving antenna. The M×1 received signal vector $r(n)$ at the input of the base station is given by:

$$r(n) = \sqrt{p_0}h_0 s_0(n) + \sum_{k=1}^{K} \sqrt{p_k}h_k s_k(n) + v(n) \qquad (15.1)$$

where $s_0(n)$ and $s_K(n)$ represents Binary Phase Shift Keying (BPSK) modulated data streams with power p_0 and p_K respectively. Here, we

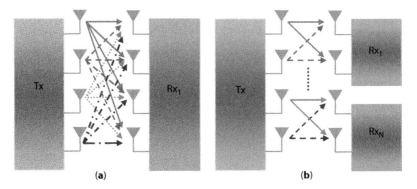

Figure 15.1 (a) Single-User system, (b) Multi-User MIMO system.

consider user 0 as the desired user and the rest as interfering users. These data bits are equiprobable and denoted as $s_0(n)$, $s_1(n)$, . . .$s_k(n) \in$ {-1, +1}. All the channel vectors $h'_k s$(k=0,1...K) are mutually independent circularly symmetric complex Gaussian M × 1 vectors such that $h_k \sim CN(0, \sigma_n^2 IM)$. The vector $v(n)$ is zero mean Gaussian noise vector such athta $v(n) \sim CN(0, \sigma_n^2 I_M)$ where I_M is an M × M identity matrix. Also, the noise is uncorrelated with interfering users data (which is true in practice). For the sake of simplicity, we drop the time index 'n' in the rest of the chapter.

We proposed a tapped delay line structure based equalizer with weights **w** as shown in Figure 15.2. The estimated signal after processing via equalizer is given by:

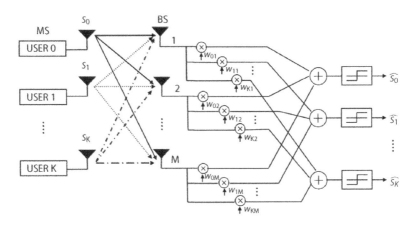

Figure 15.2 Block diagram of MU-MIMO uplink systems.

$$\widehat{s_0} = y = w_0^H * r = \sqrt{p_0}\, w_0^H h_0 s_0(n) + y_{in} \tag{15.2}$$

where $w_0 = [w_{01}\ w_{02} \ldots w_{0M}]^T$ denotes desired user's equalizer weight vector and y_{in} is the interference plus noise part after processing which is given by:

$$y_{in} = \sum_{k=1}^{K} \sqrt{p_k}\, w_0^H h_k s_k + \left(w_0^H\right) v \tag{15.3}$$

Thus, the signal-to-interference-plus-noise ratio (SINR) at the output of the receiver is given by:

$$\gamma = \frac{p_0 \left| w_0^H h_0 \right|^2}{\sigma_n^2 \|w_0\|^2 + \sum_{k=1}^{K} p_k \|w_0\|^2_{R_k}} = c \|h_0\|^2_{\widetilde{A_0}} \tag{15.4}$$

where $c = \dfrac{p_0}{\sigma_n^2 \|w_0\|^2 + \sum_{k=1}^{K} p_k \|w_0\|^2_{R_k}}$ and the weight matrix $\widetilde{A_0}$ is

defined as $\widetilde{A_0} = w_0 w_0^H$. We transform the channel h_0 to its whitened version, i.e., $\tilde{h}_0$ using the relation $\tilde{h}_0 = R_0^{-\frac{H}{2}} h_0$ and we can absorb 'c' within the weight of the quadratic norm which gives:

$$\gamma = \left\| \tilde{h}_0 \right\|^2_{cA_0} \tag{15.5}$$

where $A_0 = R_0^{\frac{1}{2}} \left(w_0 w_0^H \right) R_0^{\frac{H}{2}}$ is the weight matrix of the desired channel.

15.4   Derivation for the Probability of Error

In this section, we derive the probability of error P_e for BPSK modulated signal at the output of the equalizer. We first derive conditional P_e by conditioning on h_0 which is given by:

$$P_e(E|\gamma) = \frac{1}{2} P_e(E|h_0, s_0 = -1) + \frac{1}{2} P_e(E|h_0, s_0 = +1) \tag{15.6}$$

Since the signals (± 1) are equally probable, we can rewrite $P_e(E|\gamma)$ as

$$P_e(E|\gamma) = P_e(E|\boldsymbol{h}_0, s_0 = -1)$$

Due to condition on $\boldsymbol{h}_0$, the received statistics after processing from equalizer will be Gaussian distributed. Thus, the conditional probability of error can be obtained as

$$P_e(E|\gamma) = P_e\left(-\sqrt{p_0}\,\boldsymbol{w}_0^H \boldsymbol{h}_0 + y_{in} > 0 | \boldsymbol{h}_0\right) \tag{15.7}$$

Next, to obtain the conditional P_e in (5.7), we determine the distribution of y_{in} condition on $\boldsymbol{h}_0$. Since, $\boldsymbol{h}_k$'s are independent from $\boldsymbol{h}_0$ for k=1... K, which are distributed as $\boldsymbol{h}_k \sim CN(0, \boldsymbol{R}_k)$ and $s_k \in \{\pm 1\}$.

Thus, we have $\sqrt{p_k}\,\boldsymbol{w}_0^H \boldsymbol{h}_k s_k | \boldsymbol{h}_0, s_k \sim CN\left(0, p_k \|\boldsymbol{w}_0\|^2_{R_k} |s_k|^2\right)$. Since $|s_k|^2 = 1$, the distribution of $\sqrt{p_k}\,\boldsymbol{w}_0^H \boldsymbol{h}_k s_k | \boldsymbol{h}_0$ conditioned on s_k is the equivalence to the distribution of $\sqrt{p_k}\,\boldsymbol{w}_0^H \boldsymbol{h}_k s_k | \boldsymbol{h}_0$ itself. i.e.,

$$\sqrt{p_k}\,\boldsymbol{w}_0^H \boldsymbol{h}_k s_k | \boldsymbol{h}_0 \sim CN\left(0, p_k \|\boldsymbol{w}_0\|^2_{R_k}\right) \text{ for } k = 1, \ldots, K \tag{15.8}$$

As a result, y_{in} condition on $\boldsymbol{h}_0$ is distributed as:

$$y_{in} = CN(0, \sigma^2) \tag{15.9}$$

where $\sigma^2 = \sigma_n^2 \|\boldsymbol{w}_0\|^2 + \sum_{k=1}^{K} p_k \|\boldsymbol{w}_0\|^2_{R_k}$ is the variance of y_{in}

Using the well-known ML estimation [26] to evaluate the probability of error $P_e(E|\gamma) = Q\left(\dfrac{\|\text{v}_A - \text{v}_B\|}{\sqrt{2\sigma^2}}\right)$ where $\text{v}_A = \sqrt{p_0}\,\boldsymbol{w}_0^H \boldsymbol{h}_0$ when s0=+1 and $\text{v}_B = -\sqrt{p_0}\,\boldsymbol{w}_0^H \boldsymbol{h}_0$ when s0=-1, the conditional probability of error $P_e(\gamma|\boldsymbol{h}_0)$ is found to be:

$$P_e(E|\gamma) = Q\left(\sqrt{2\gamma}\right) \tag{15.10}$$

Now, to obtain the probability of error, we must average the conditional probability of error in (5.10) with the PDF of SINR which is expressed as:

$$P_e(E) = \int_0^\infty P_e(E|\gamma) f\gamma(\gamma) d\gamma \tag{15.11}$$

To proceed further, we require the PDF of SINR using (5.5), which is in indefinite quadratic form. Now by using [26], we get the PDF of SINR in the closed form as:

$$f_\gamma(\gamma) = \frac{1}{|c|} \sum_{t=1}^{M} \frac{\lambda_t^{M-1}}{|\lambda_t| \prod\limits_{j=1, j \neq t}^{M} (\lambda_t - \lambda_j)} e^{-\frac{\gamma}{C\lambda_t}} u\left(\frac{\gamma}{C\lambda_t}\right) \quad (15.12)$$

where λ's are the eigenvalues of A_0 obtained as:

$$A_0 = Q \wedge Q^H \quad (15.13)$$

Substituting the PDF of SINR from equation (5.12) in (5.11), we get:

$$P_e(E) = \frac{1}{|c|} \sum_{t=1}^{M} \frac{\lambda_t^{M-1}}{|\lambda_t| \prod\limits_{j=1, j \neq t}^{M} (\lambda_t - \lambda_j)} \int_0^\infty Q(\sqrt{2\gamma}) e^{-\frac{\gamma}{C\lambda_t}} d\gamma \quad (15.14)$$

If, we represent the Q-function in terms of the error function:

$$Q(x) = \frac{1}{2}\left(1 - erf\left(\frac{x}{\sqrt{2}}\right)\right)$$

we can transform (5.14) as:

$$P_e(E) = \frac{1}{2|c|} \sum_{t=1}^{M} \frac{\lambda_t^{M-1}}{|\lambda_t| \prod\limits_{j=1, j \neq t}^{M} (\lambda_t - \lambda_j)} \int_0^\infty \left(1 - erf\left(\sqrt{\gamma}\right)\right) e^{-\frac{\gamma}{C\lambda_t}} d\gamma \quad (15.15)$$

Integration in terms of the error function and the exponential term, from [10] gives:

$$\int_0^\infty erf\left(\sqrt{qt}\right) e^{-Pt} dt = \frac{\sqrt{q}}{p} \frac{1}{\sqrt{P+q}} \quad \text{when } Re: p > 0 \, \& \, Re: q + p > 0 \quad (15.16)$$

The above integration (16) can be evaluated using the above result. Thus, the $P_e(\gamma)$ is found to be:

$$P_e(E) = \frac{1}{2|c|} \sum_{t=1}^{M} \frac{\lambda_t^{M-1}}{|\lambda_t| \prod_{j=1, j \neq t}^{M} (\lambda_t - \lambda_j)} \left[c\lambda_t \left(1 - \frac{1}{\sqrt{\frac{1}{c\lambda_t} + 1}} \right) \right] \quad (15.17)$$

15.5   Design of Adaptive Equalizer by Minimizing BER

In the proposed work, we design the equalizer that minimizes the probability of error derived in section 15.3 with the constrained of unity norm of the weight vector, i.e.,

$$\underset{w_0}{\text{minimize}} \, P_e(E)$$
$$subject\ to \, \|w_0\|^2 = 1 \quad (15.18)$$

To achieve the above goal, we utilize two approaches: Interior Point Approach and Stochastic Gradient Approach.

15.5.1   Interior Point Approach

The problem of convex optimization for linear and Nonlinear can be solved using an Interior Point Algorithm. It is also known as Barrier Method [27]. We use the built-in tool in Matlab to implement this technique.

15.5.2   Stochastic Gradient Approach

To acquire the minimum probability of error, an adaptive equalizer is implemented using a stochastic gradient approach.

$$w_0(n+1) = w_0(n) - \mu \frac{\partial P_e(E)}{\partial w_0} \quad (15.19)$$

where μ represent the step size and $P_e(E)$ is an objective function. To present the concept of stochastic gradient based equalizer design we consider a special case of M=2. For this special case, the probability of error can be represented as:

$$P_e(E) = \frac{c}{2\,|c|} \sum_{t=1}^{2} \frac{\lambda_t^2}{|\lambda_t|(\lambda_t - \lambda_j)} - \frac{c^{\frac{3}{2}}}{2\,|c|} \sum_{t=1}^{2} \frac{\lambda_t^{\frac{5}{2}}}{|\lambda_t|(\lambda_t - \lambda_j)\sqrt{1 + c\lambda_t}} \quad (15.20)$$

We differentiate the above $P_e(E)$ with respect to $\mathbf{w}_0$ (for M=2, $\mathbf{w}_0 = [\,\mathbf{w}_{01}\ \mathbf{w}_{02}\,]^T$). Thus, we need to find $\dfrac{\partial P_e(E)}{\partial w_{0i}}$, where i=1,2. Note that the any eigenvalue λ_n can be set up as $\lambda_n = \|q_n\|_{A_0}^2$ where q_n is the nth eigenvector of matrix Q in (15.13) affecting in (15.19) are a function of $\mathbf{w}_0$. Therefore, we need to evaluate $\dfrac{\partial \lambda_n}{\partial w_{0i}} = \|q_n\|^2 \, {}_{\!\!R_0^{\frac{H}{2}}(G_i)R_0^{\frac{1}{2}}}$ whereby using the complex derivative, $\mathbf{G}_i = \dfrac{\partial(\mathbf{w}_0 \mathbf{w}_0^H)}{\partial w_{0i}}$ is a 2×2 matrix with all zeros except the ith row equals to $\mathbf{w}_0^H$.

To proceed further, we evaluate $\dfrac{\partial c}{\partial w_{0i}}$ and $\dfrac{\partial |c|}{\partial w_{0i}}$ as:

$$c_{0i} = \frac{\partial c}{\partial w_{0i}} = \frac{-p_0 \left[\sigma_n^2 w_{0i}^* + \sum_{k=1}^{K} p_k \lambda_k w_{01}^* \right]}{\left[\sigma_n^2 \|\mathbf{w}_0\|^2 + \sum_{k=1}^{K} p_k \|\mathbf{w}_0\|_{R_k}^2 \right]^2} \quad (15.21)$$

$$P_i = \frac{\partial |c|}{\partial w_{0i}} = sign(c)c_{0i} \quad (15.22)$$

similarly by using $\dfrac{\partial \lambda_n}{\partial w_{0i}}$, we find $\dfrac{\partial |\lambda_t|}{\partial w_{0i}}$, $\dfrac{\partial(\lambda_t - \lambda_j)}{\partial w_{0i}}$ and $\dfrac{\partial(\sqrt{1 + c\lambda_t})}{\partial w_{0i}}$ as follows:

$$Q_i = \frac{\partial |\lambda_t|}{\partial w_{0i}} = sign(\lambda_t) \|q_t\|^2 \, {}_{\!\!R_0^{\frac{H}{2}}(G_i)R_0^{\frac{1}{2}}} \quad (15.23)$$

$$R_i = \frac{\partial(\lambda_t - \lambda_j)}{\partial w_{0i}} = \|q_t\|^2 \, {}_{\!\!R_0^{\frac{H}{2}}(G_i)R_0^{\frac{1}{2}}} - \|q_j\|^2 \, {}_{\!\!R_0^{\frac{H}{2}}(G_i)R_0^{\frac{1}{2}}} \quad (15.24)$$

$$S_i = \frac{\partial\left(\sqrt{1+c\lambda_t}\right)}{\partial w_{0i}} = \frac{1}{2\sqrt{1+c\lambda_t}}\left[c\,\|q_t\|^2_{\substack{H \\ R_0^{\frac{1}{2}}(G_i)R_0^{\frac{1}{2}}}} + \lambda_t c_{0i}\right] \quad (15.25)$$

The $\dfrac{\partial P_e(\gamma)}{\partial w_{0i}}$ is a difference of two differentiation terms namely I_{1i} and I_{2i}. The first differentiation term I_{1i} is simplified as:

$$I_{1i} = \frac{\partial\left(\dfrac{c}{2|c|}\sum_{t=1}^{2}\dfrac{\lambda_t^2}{|\lambda_t|(\lambda_t - \lambda_j)}\right)}{\partial w_{0i}} = \frac{1}{2}\sum_{t=1}^{2}\left[\frac{|c||\lambda_t|(\lambda_t - \lambda_j)U_{ni} - c\lambda_t^2 U_{di}}{(|c||\lambda_t|(\lambda_t - \lambda_j))^2}\right]$$

$$(15.26)$$

Where

$$U_{ni} = \frac{\partial\left(c\lambda_t^2\right)}{\partial w_{0i}} = 2c\lambda_t\,\|q_t\|^2_{\substack{H \\ R_0^{\frac{1}{2}}(G_i)R_0^{\frac{1}{2}}}} + \lambda_t^2 c_{0i} \quad (15.27)$$

and

$$U_{di} = \frac{\partial\left(|c||\lambda_t|(\lambda_t - \lambda_j)\right)}{\partial w_{0i}} \quad (15.28)$$

$$= P_i\,|\lambda_t|(\lambda_t - \lambda_j) + Q_i\,|c|(\lambda_t - \lambda_j) + R_i\,|c||\lambda_t|$$

similarly, the second differentiation term $I2i$ is simplified as

$$I_{2i} = \frac{1}{2}\sum_{t=1}^{2}\left[\frac{\left(|c||\lambda_t|(\lambda_t - \lambda_k)\sqrt{1+c\lambda_t}\right)V_{ni} - \left(\lambda_t^{\frac{5}{2}}c^{\frac{3}{2}}V_{di}\right)}{\left(|c||\lambda_t|(\lambda_t - \lambda_j)\sqrt{1+c\lambda_t}\right)^2}\right] \quad (15.29)$$

Where

$$V_{ni} = \frac{\partial c^{\frac{3}{2}}.\lambda_t^{\frac{5}{2}}}{\partial\,w_{0i}} = \frac{3}{2}\lambda_t^{\frac{5}{2}}c^{\frac{1}{2}}c_{0i} + \frac{5}{2}\lambda_t^{\frac{3}{2}}c^{\frac{3}{2}}\,\|qt\|^2_{\substack{H \\ R_0^{\frac{1}{2}}(G_i)R_0^{\frac{1}{2}}}} \quad (15.30)$$

and

$$V_{di} = \frac{\partial\left(|c||\lambda_t|(\lambda_t - \lambda_j)\sqrt{1+c\lambda_t}\right)}{\partial w_{0i}}$$

$$= P_i\,|\lambda_t|\,(\lambda_t - \lambda_j)\sqrt{1+c\lambda_t} + Q_i\,|c|\,(\lambda_t - \lambda_j)\sqrt{1+c\lambda_t} + R_i\,|c||\lambda_t|\,\sqrt{1+c\lambda_t}$$

$$+ S_i\,|c|\,(\lambda_t - \lambda_j)|\lambda_t|$$

15.6   Simulation Results

In this section, we compare the probability of error for Maximum Ratio Combining (MRC) [28] with our proposed equalizer in Figure 15.3. Since it is a well-known fact that MRC outperforms MMSE and ZF equalizer [29], we use the probability of error using MRC as a benchmark for our work. However, we consider the scenario of both perfect and imperfect channel knowledge in the implementation of MRC. We assume that channel has a random error with variance σ_e^2. Here we consider different values of σ_e^2 which are 0 (perfect channel knowledge), 0.3, 0.5, 1.0, 1.5 etc. We observe in the figure that the performance of MRC is better than ours. However, its performance degrades as the uncertainty in the channel increases.

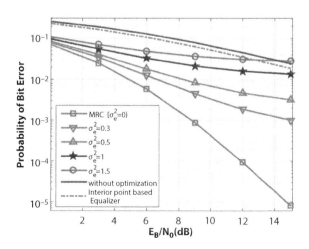

Figure 15.3 Comparison of the probability of error using interior point based equalizer and MRC with different level of uncertainty in the channel knowledge.

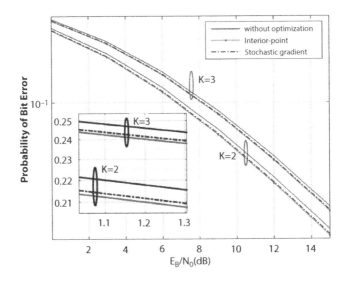

Figure 15.4 Comparison of the probability of error perform over Interior point and stochastic gradient equalizer for M=2.

We also investigate the effect of different optimization techniques where K is the number of users, and M is the number of antenna elements. In Figure 15.4, we compare the performance of probability of error using a stochastic gradient and interior point approach for two different numbers of users, i.e., K=2, 3, and M=2. It shows that the interior point approach has better performance in contrast to the stochastic gradient approach.

15.7   Summary

We have implemented an adaptive equalizer to minimize the probability of error for MU-MIMO uplink system. This design is dealt with BPSK modulation scheme without using CSI at the receiver. We derived closed form expression for the probability of error of the BPSK modulated system, which is utilized in the development of the equalizer. An Adaptive equalizer is designed using two different methods: Interior point and stochastic gradient. It is seen that the interior point method has better performance in contrast to the stochastic gradient approach. In the future, this work can be further elaborated for QPSK, 8PSK, and 16PSK, etc.

References

1. E. Telatar, "Capacity of Multi-Antenna Gaussian channels", *AT&T Bell Labs. Tech. Memo.*, vol. 10, no. 6, pp. 585–595, Oct. 1995.
2. Q. H. Spencer, C. B. Peel, A. L. Swindlehurst, and M. Haardt, "An introduction to the multiuser MIMO downlink," *IEEE communications Magazine*, vol. 42, pp. 60-67, 2004.
3. V. Stankovic and M. Haardt, "Novel linear and nonlinear multiuser MIMO downlink precoding with improved diversity and capacity," in *Proc. of the 16th Meeting of the Wireless World Research Forum (WWRF), Shanghai, China*, 2006.
4. F. Farrokhi, G. Foschini, A. Lozano, and R. Valenzuela, "Link-optimal BLAST processing with multiple-access interference," in *Vehicular Technology Conference, 2000. IEEE-VTS Fall VTC 2000. 52nd*, 2000, pp. 87-91.
5. R. S. Blum, "MIMO capacity with interference," *IEEE Journal on selected areas in communications*, vol. 21, pp. 793-801, 2003.
6. R. W. Heath, M. Airy, and A. J. Paulraj, "Multiuser diversity for MIMO wireless systems with linear receivers," in *Signals, Systems and Computers, 2001. Conference Record of the Thirty-Fifth Asilomar Conference on*, 2001, pp. 1194-1199.
7. R. Knopp and P. A. Humblet, "Information capacity and power control in single-cell multiuser communications," in *Communications, 1995. ICC'95 Seattle,' Gateway to Globalization', 1995 IEEE International Conference on*, 1995, pp. 331-335.
8. E. G. Larsson, O. Edfors, F. Tufvesson, and T. L. Marzetta, "Massive MIMO for next generation wireless systems," *IEEE Communications Magazine*, vol. 52, pp. 186-195, 2014.
9. M. Joham, W. Utschick, and J. A. Nossek, "Linear transmit processing in MIMO communications systems," *IEEE Transactions on signal Processing*, vol. 53, pp. 2700-2712, 2005.
10. A. Mueller, A. Kammoun, E. Björnson, and M. Debbah, "Linear precoding based on polynomial expansion: Reducing complexity in massive MIMO," *EURASIP Journal on Wireless Communications and Networking*, vol. 2016, p. 63, 2016.
11. Q. H. Spencer and M. Haardt, "Capacity and downlink transmission algorithms for a multiuser MIMO channel," in *Signals, Systems and Computers, 2002. Conference Record of the Thirty-Sixth Asilomar Conference on*, 2002, pp. 1384-1388.
12. K. Zu, R. C. de Lamare, and M. Haardt, "Generalized Design of Low-Complexity Block Diagonalization Type Precoding Algorithms for Multiuser MIMO Systems," *IEEE Trans. Communications*, vol. 61, pp. 4232-4242, 2013.
13. V. Stankovic and M. Haardt, "Multiuser MIMO downlink precoding for users with multiple antennas," in *Proc. of the 12-th Meeting of the Wireless World Research Forum (WWRF), Toronto, ON, Canada*, 2004, pp. 12-14.
14. W. Ni and X. Dong, "Hybrid block diagonalization for massive multiuser MIMO systems," *IEEE transactions on communications*, vol. 64, pp. 201-211, 2016.

15. C. Masouros, M. Sellathurai, and T. Ratnarajah, "Maximizing energy efficiency in the vector precoded MU-MISO downlink by selective perturbation," *IEEE Transactions on Wireless Communications,* vol. 13, pp. 4974-4984, 2014.

16. M. Mazrouei-Sebdani, W. A. Krzymień, and J. Melzer, "Massive MIMO With Nonlinear Precoding: Large-System Analysis," *IEEE Transactions on Vehicular Technology,* vol. 65, pp. 2815-2820, 2016.

17. S. Mydhili and A. Rajeswari, "Equalization techniques in MIMO systems-An analysis," in *Communications and Signal Processing (ICCSP), 2014 International Conference on,* 2014, pp. 1082-1086.

18. H. Simon, *Adaptive filter theory,* Prentice Hall, vol. 2, pp. 478-481, 2002.

19. A. H. Mehana and A. Nosratinia, "Performance of linear receivers in frequency-selective MIMO channels," *IEEE Transactions on Wireless Communications,* vol. 12, pp. 2697-2705, 2013.

20. A. Agarwal, S. Sur, A. K. Singh, H. Gurung, A. K. Gupta, and R. Bera, "Performance analysis of linear and nonlinear equalizer in Rician channel," *Procedia Technology,* vol. 4, pp. 687-691, 2012.

21. D. Mattera, F. Palmieri, and G. D'Angelo, "Bidirectional MIMO equalizer design," *in Signal Processing Conference, 2006* 14th European, 2006, pp. 1-5.

22. G. M. Giivensen and A. O. Yilmaz, "Iterative frequency domain equalization for single-carrier wideband MIMO channels," in *Personal, Indoor and Mobile Radio Communications, 2009 IEEE 20th International Symposium on,* 2009, pp. 2661-2665.

23. M. Aaron and D. Tufts, "Intersymbol interference and error probability," *IEEE Transactions on Information Theory,* vol. 12, pp. 26-34, 1966.

24. E. Shamash and K. Yao, "On the structure and performance of a linear decision feedback equalizer based on the minimum error probability criterion," in *International Conference on Communications, 10th, Minneapolis, Minn,* 1974, p. 25.

25. S. Chen, E. Chng, B. Mulgrew, and G. Gibson, "Minimum-BER linear-combiner DFE," in *Communications, 1996. ICC'96, Conference Record, Converging Technologies for Tomorrow's Applications. 1996 IEEE International Conference on, 1996,* pp. 1173-1177.

26. T. Y. Al-Naffouri, M. Moinuddin, N. Ajeeb, B. Hassibi, and A. L. Moustakas, "On the distribution of indefinite quadratic forms in Gaussian random variables," *IEEE Transactions on Communications,* vol. 64, pp. 153-165, 2016.

27. G. B. Dantzig and M. N. Thapa, *Linear programming 2: theory and extensions:* Springer Science & Business Media, 2006.

28. A. A. Basri and T. J. Lim, "Exact average bit-error probability for maximal ratio combining with multiple co-channel interferers and Rayleigh fading," in *Communications, 2007. ICC'07. IEEE International Conference on, 2007,* pp. 1102-1107.

29. J. R. Barry, E. A. Lee, and D. G. Messerschmitt, *Digital communication:* Springer Science & Business Media, 2012.

About the Editors

Prashant Ranjan, PhD, is an associate professor in the Department of Electronics and Communication Engineering, University of Engineering and Management Jaipur, Rajasthan, India. He earned his masters and doctorate from Motilal Nehru National Institute of Technology Allahabad, Prayagraj, Uttar Pradesh, India. He has more than 4 years of teaching experience and has published numerous research papers in international journals and conferences. He has also served as a reviewer for a number of technical journals and conferences.

Dharmendra Kumar Jhariya, PhD, is an assistant professor in the Department of Electronics and Communication, National Institute of Technology Delhi, India. He earned his doctorate from the Indian Institute of Technology, Kharagpur. He has more than five years of teaching experience and has published numerous research papers in international scientific journals and conferences.

Manoj Gupta, PhD, is an associate professor in the Department of Electronics and Communication Engineering, JECRC University, Jaipur (Rajasthan), India. Earned his doctorate from the University of Rajasthan, Jaipur, India. He has over fifteen years of teaching experience and has published many research papers in scientific journals and conferences. He has contributed numerous book chapters to edited volumes and has four patents to his credit. He is the Editor in Chief of the book series "Advances in Antenna, Microwave and Communication Engineering," from Scrivener Publishing, and he is editor in chief of a book series by another publisher. He has spoken at and been involved in numerous scientific conferences and was the keynote speaker at the 2017 IEEE International Conference on Signal and Image Processing and at the 2017 International Conferences on Public Health and Medical Sciences in Xi'an, China. He is an editor, associate editor, and reviewer for many international technical journals and has received numerous awards. He is listed in *Marquis Who's Who in Science and Engineering*® USA and *Marquis Who's Who in the World*® USA.

Er. Krishna Kumar is a research and development engineer at UJVN Limited and is pursuing his PhD from the Indian Institute of Technology, Roorkee. He has more than eleven years of experience in this field and has published numerous research papers in international journals and conferences from well-respected publishers.

Pradeep Kumar, PhD, has over fourteen years of teaching experience and is working with the University of KwqZulu-Natal, South Africa. He is the co-editor in chief of the book series "Advances in Antenna, Microwave, and Communication Engineering," from Scrivener Publishing and has received numerous awards and fellowships. He is also the author of more than 90 research papers published in various peer-reviewed scientific journals and conferences and a reviewer for many journals and conferences. He is also an academic member of the Center of Excellence, University of KwaZulu-Natal.

Index